THE HISTORY *of* MEDICINE

MEDICINE TODAY

2000 TO THE PRESENT

THE HISTORY of MEDICINE

MEDICINE TODAY

2000 TO THE PRESENT

KATE KELLY

Facts On File
An imprint of Infobase Publishing

MEDICINE TODAY: 2000 to the Present

Facts On File, Inc.
An imprint of Infobase Publishing
132 West 31st Street
New York NY 10001

Library of Congress Cataloging-in-Publication Data

Kelly, Kate, 1950–
   Medicine today: 2000 to the present / Kate Kelly.
     p. ; cm.—(History of medicine)
   Includes bibliographical references and index.
   ISBN 978-0-8160-7210-1 (alk. paper)
   1. Medicine—History—21st century. I. Title. II. Series: History of medicine (Facts on File, Inc.)
   [DNLM: I. History of Medicine. 2. History, 21st Century. 3. Ethics, Medical. 4. History, 20th Century. WZ40 K29ma 2010]
   R131.K43 2010
   610.9—dc22   2009016629

Facts On File books are available at special discounts when purchased in bulk quantities for businesses, associations, institutions, or sales promotions. Please call our Special Sales Department in New York at (212) 967-8800 or (800) 322-8755.

You can find Facts On File on the World Wide Web at http://www.factsonfile.com

Excerpts included herewith have been reprinted by permission of the copyright holders; the author has made every effort to contact copyright holders. The publishers will be glad to rectify, in future editions, any errors or omissions brought to their notice.

Text design by Annie O'Donnell
Illustrations by Bobbi McCutcheon
Photo research by Elizabeth H. Oakes
Composition by Hermitage Publishing Services
Cover printed by Bang Printing, Inc., Brainerd, Minn.
Book printed and bound by Bang Printing, Inc., Brainerd, Minn.
Date printed: February, 2010
Printed in the United States of America

10 9 8 7 6 5 4 3 2 1

This book is printed on acid-free paper.

CONTENTS

> *"You have to know the past to understand the present."*
> —*American scientist Carl Sagan (1934–96)*

The history of medicine offers a fascinating lens through which to view humankind. Maintaining good health, overcoming disease, and caring for wounds and broken bones was as important to primitive people as it is to us today, and every civilization participated in efforts to keep its population healthy. As scientists continue to study the past, they are finding more and more information about how early civilizations coped with health problems, and they are gaining greater understanding of how health practitioners in earlier times made their discoveries. This information contributes to our understanding today of the science of medicine and healing.

In many ways, medicine is a very young science. Until the mid-19th century, no one knew of the existence of germs, so as a result, any solutions that healers might have tried could not address the root cause of many illnesses. Yet for several thousand years, medicine has been practiced, often quite successfully. While progress in any field is never linear (very early, nothing was written down; later, it may have been written down, but there was little intra-community communication), readers will see that some civilizations made great advances in certain health-related areas only to see the knowledge forgotten or ignored after the civilization faded. Two early examples of this are Hippocrates' patient-centered healing philosophy and the amazing contributions of the Romans to public health through water-delivery and waste-removal systems. This knowledge was lost and had to be regained later.

The six volumes in the History of Medicine set are written to stand alone, but combined, the set presents the entire sweep of the history of medicine. It is written to put into perspective for high school students and the general public how and when various medical discoveries were made and how that information affected health care of the time period. The set starts with primitive humans and concludes with a final volume that presents readers with the very vital information they will need as they must answer society's questions of the future about everything from understanding one's personal risk of certain diseases to the ethics of organ transplants and the increasingly complex questions about preservation of life.

Each volume is interdisciplinary, blending discussions of the history, biology, chemistry, medicine and economic issues and public policy that are associated with each topic. *Early Civilizations,* the first volume, presents new research about very old cultures because modern technology has yielded new information on the study of ancient civilizations. The healing practices of primitive humans and of the ancient civilizations in India and China are outlined, and this volume describes the many contributions of the Greeks and Romans, including Hippocrates' patient-centric approach to illness and how the Romans improved public health.

The Middle Ages addresses the religious influence on the practice of medicine and the eventual growth of universities that provided a medical education. During the Middle Ages, sanitation became a major issue, and necessity eventually drove improvements to public health. Women also made contributions to the medical field during this time. *The Middle Ages* describes the manner in which medieval society coped with the Black Death (bubonic plague) and leprosy, as illustrative of the medical thinking of this era. The volume concludes with information on the golden age of Islamic medicine, during which considerable medical progress was made.

The Scientific Revolution and Medicine describes how disease flourished because of an increase in population, and the book describes the numerous discoveries that were an important aspect of this time. The volume explains the progress made by Andreas Vesalius (1514–64) who transformed Western concepts of the structure of the human body; William Harvey (1578–1657), who studied and wrote about the circulation of the human blood; and Ambroise Paré (1510–90), who was a leader in surgery. Syphilis was a major scourge of this time, and the way that society coped with what seemed to be a new illness is explained. Not all beliefs of this time were progressive, and the occult sciences of astrology and alchemy were an important influence in medicine, despite scientific advances.

Old World and New describes what was happening in the colonies as America was being settled and examines the illnesses that beset them and the way in which they were treated. However, before leaving the Old World, there are several important figures who will be introduced: Thomas Sydenham (1624–89) who was known as the English Hippocrates, Herman Boerhaave (1668–1738) who revitalized the teaching of clinical medicine, and Johann Peter Frank (1745–1821) who was an early proponent of the public health movement.

Medicine Becomes a Science begins during the era in which scientists discovered that bacteria was the cause of illness. Until 150 years ago, scientists had no idea why people became ill. This volume describes the evolution of "germ theory" and describes advances that followed quickly after bacteria was identified, including vaccinations, antibiotics, and an understanding of the importance of cleanliness. Evidence-based medicine is introduced as are medical discoveries from the battlefield.

Medicine Today examines the current state of medicine and reflects how DNA, genetic testing, nanotechnology, and stem cell research all hold the promise of enormous developments within the course of the next few years. It provides a framework for teachers and students to understand better the news stories that are sure to be written on these various topics: What are stem cells, and why is investigating them so important to scientists? And what is nanotechnology? Should genetic testing be permitted? Each of the issues discussed are placed in context of the ethical issues surrounding it.

Each volume within the History of Medicine set includes an index, a chronology of notable events, a glossary of significant terms and concepts, a helpful list of Internet resources, and an array of historical and current print sources for further research. Photographs, tables, and line art accompany the text.

I am a science and medical writer with the good fortune to be assigned this set. For a number of years I have written books in collaboration with physicians who wanted to share their medical knowledge with laypeople, and this has provided an excellent background in understanding the science and medicine of good health. In addition, I am a frequent guest at middle and high schools and at public libraries addressing audiences on the history of U.S. presidential election days, and this regular experience with students keeps me fresh when it comes to understanding how best to convey information to these audiences.

What is happening in the world of medicine and health technology today may affect the career choices of many, and it will affect the health care of all, so the topics are of vital importance. In addition, the public health policies under consideration (what medicines to develop, whether to permit stem cell research, what health records to put online, and how and when to use what types of technology, etc.) will have a big impact on all people in the future. These subjects are in the news daily, and students who can turn to authoritative science volumes on the topic will be better prepared to understand the story behind the news.

ACKNOWLEDGMENTS

This book as well as the others in the set was made possible because of the guidance, inspiration, and advice offered by many generous individuals who have helped me better understand science and medicine and their histories. I would like to express my heartfelt appreciation to Frank Darmstadt, whose vision and enthusiastic encouragement, patience, and support helped shape the series and see it through to completion. Thank you, too, to the Facts On File staff members who worked on this set.

The line art and the photographs for the entire set were provided by two very helpful professionals—the artist Bobbi McCutcheon provided all the line art; she frequently reached out to me from her office in Juneau, Alaska, to offer very welcome advice and support as we worked through the complexities of the renderings. A very warm thank you to Elizabeth Oakes for finding a wealth of wonderful photographs that helped bring the information to life. Carol Sailors got me off to a great start, and Carole Johnson kept me sane by providing able help on the back matter of all the books. Agent Bob Diforio has remained steadfast in his shepherding of the work.

I also want to acknowledge the wonderful archive collections that have provided information for the book. Without places like the Sophia Smith Collection at the Smith College library, firsthand accounts of the Civil War battlefield treatment or reports like Lillian Gilbreth's on helping the disabled after World War I would be lost to history.

"If I have seen further it is by standing on the shoulders of giants."

Isaac Newton, letter to Robert Hooke, February 5, 1675

People have managed health issues since the beginning of time, and in the process they have unraveled a great deal of information about anatomy, physiology, and mental health. Our medical progress today must be attributed to the work previously done by scientists and physicians who noted what worked and almost as important what ideas failed to produce the results they hoped for. Amazingly, it has only been in the last 160 years that scientists have known that bacteria and viruses (germs) cause illness, so the future of important discoveries in this field is still very much wide open. Today's young people are at the forefront of a field where there will be great progress in finding ways to keep people healthier and help them live for a very long time.

Today's world of medicine is filled with infinite possibilities as well as many moral quandaries that attend any type of great progress. Laypeople sometimes assume that physicians are fully prepared to offer guidance on difficult medical issues, but medicine is both a science and an art, and physicians are faced with perplexing decisions almost every day as they try to navigate the best treatment for each person's condition. Scientists in the laboratory face difficult issues too. Just because it is possible to harvest stem cells does not mean they should be harvested; the same issue arises with cloning. While one day nanotechnology might mean a little nanobot can be inserted into a blood vessel to clean out an artery, there are some ethical and environmental issues that need to be considered first. Then there are the issues surrounding pregnancies and births—everything from birth control to abortions can be a hot button issue for some group of people or another.

Today birth defects can be diagnosed while a baby is still in utero, putting parents in very fraught positions. These are difficult and emotional issues with no easy answers, and everyone is working toward figuring out how to do what is right.

The Internet presents medicine with wonderful opportunities and real risks. Easy access to personal health information online can be a tremendous benefit for the sick person who arrives unconscious in an emergency room, but this type of online access to medical records leads to privacy issues. There is always the threat that the information could be accessed by a current or future employer who will determine that the person's condition makes them risky to employ. Web surfing by patients is also a blessing and a curse. On one hand, physicians find that patients are much more knowledgeable because of the research they do on their own conditions, but patients lack the background to fully understand the range of possibilities and often become alarmed over nothing. This creates new challenges for physicians.

Today's scientists know definitively that some of the illnesses people suffer today are caused by the environment—mercury in our food supply, asbestos contamination in buildings, and industrial pollutants in our air that lead to illnesses ranging from asthma to lung cancer. As countries become more densely settled, pollution and access to freshwater becomes increasingly difficult. Scientists are also learning that a good number of our diseases start out in the animal population. The careful tracking of swine (H1N1) and avian flu—still primarily in the bird population—is an excellent example of both scientists' concerns and the types of measures that can be put in place to try to halt diseases before they jump species. Government leaders, scientists, and laypeople alike worry about what measures should be taken to protect populations.

Medicine Today helps readers focus on ways to ask the right questions about these and other issues. Chapter 1 sets the scene for the book by explaining the purpose of the Centers for Disease Control and Prevention and highlighting some medical advances such as organ transplants that are taken for granted but are actually very recent developments. Chapter 2 examines the link between health and the environment. Mercury exposure, lead levels, asbestos, and radon are just a few of the pollutants to which modern people are exposed, and readers will gain a better understanding of whether regulation of these issues can be effective. Chapter 3 discusses conservation medicine. As man increasingly encroaches on animal habitats, one of the results has been an increase in cross-species transmission of disease. Physicians and veterinarians around the world are all taking serious stock of what is happening in order to reduce illness among both humans and animals. Chapter 4 examines medical frontiers of the future, addressing stem cell research and both the controversy and the promise of it. Also discussed are genetic medicine and nanotechnology. These three fields are exciting but they are not risk free.

Chapter 5 exposes the prickly subject of medical ethics. How do medical professionals reconcile what they can do with what they should do? A physician can resuscitate a 92-year-old with major health problems, but most people believe that we need to think about these quality of life issues. Chapter 6 takes on the difficult topic of paying for health care. Anyone who has followed politics in the last 20 years knows that countries need to come up with solutions on how to provide affordable health care for the most people possible. Chapter 7 discusses preventive medical measures that are being studied by physicians and public health departments. Now people are starting to realize encouraging people to lose weight and eat right and finding ways to encourage industry to keep the air clean are health issues. Chapter 8 is a view of medicine of the future including telemedicine and surgery by remote control. The possibilities of the future are both amazing and enticing.

Medicine Today takes a fascinating look at the world of medicine today with a focus on the possibilities of tomorrow. The back matter of the volume contains a chronology, a glossary, and an array of historical and current sources for further research. These sections should prove especially helpful for readers who need additional information on specific terms, topics, and developments in medical science.

This book is a vital addition to literature on the history of medicine because it puts into perspective the medical discoveries of the period and provides readers with a better understanding of the current state of medicine. Young people of today will experience a world of medicine that is going to keep evolving, and the primary purpose of this volume will be to provide them with the background necessary to keep asking the right questions about medical decisions of the future.

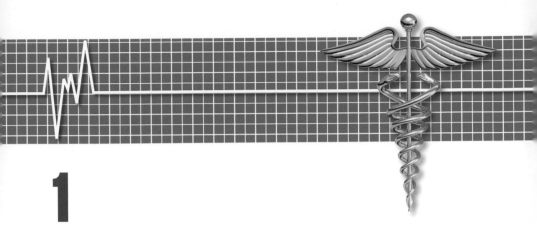

1

Recently Crossed Medical Frontiers

To anticipate the future, a society must understand the recent past. Today, newspaper stories, television reports, and Webcasts talk extensively about organ transplants, *in vitro fertilization*, and amazing robotic types of surgery. These issues are reported as perfectly normal occurrences. To provide perspective for the future, it is enlightening to realize how very recently some of these possibilities have become a reality.

In addition to all types of new scientific developments, an element that distinguishes health care today from health care of earlier centuries is that most countries have a nationally organized entity to oversee health issues that affect the citizens. The emphasis ranges from guidelines about the necessary *vaccines* for travel or for children to overseeing programs like Medicare and Medicaid. In the United States, the overall umbrella that encompasses all health-related matters is the Department of Health and Human Services (HHS). While the HHS has many tentacles, there are two parts of it that are most commonly in the public eye and this chapter will explain the purpose of both of them—the Centers for Disease Control and Prevention (CDC) and the Office of the Surgeon General (OSG).

CDC headquarters in Atlanta *(Centers for Disease Control and Prevention)*

This chapter will offer two perspectives. It will introduce the U.S. governmental mechanism that is intended to provide oversight over the health of its citizens, and it will describe some of the recent advances that have led to amazing changes in medicine. There have been so many notable medical achievements in recent years that it is difficult to select just a few on which to focus. However, organ transplants, in vitro fertilization, which was highly controversial only 40 years ago, and some of the new types of robotic surgery are certainly worthy of special note.

THE CENTERS FOR DISEASE CONTROL AND PREVENTION

A farsighted physician Dr. Joseph W. Mountin (1891–1952) founded the entity now known as the Centers for Disease Control and Prevention in 1946. The idea for the center grew from the existence of a bureau that primarily focused on controlling malaria outbreaks in the southern United States during World War II. Malaria was endemic to the southeastern United States, and during the 1940s when troops were arriving at military bases in the area to prepare for departure for Europe, the government formed a health bureau based in Atlanta, Georgia, to try to control the outbreaks of malaria affecting the men. At that time, the focus of the program was preventive and primarily involved the use of DDT, which over time successfully eliminated both malaria and typhus in the United States. Mountin wanted to expand the bureau so that it could operate on a national level and offer aid to any state facing any health crisis. Mountin's plans and vision for this bureau, which was to evolve into the CDC, is why this government entity is based in Atlanta.

The study of disease, known as *epidemiology,* was a small specialty in the mid-20th century, but Mountin intended to create a world-class center for epidemiology. In 1949, he hired Dr. Alexander Langmuir (1910–93) to head the organization. As Langmuir settled into his job, one issue he noted was that the government could reduce spending on malaria prevention; the disease had all but disappeared in the United States. By 1951, what was then called the Communicable Disease Center began to shift emphasis from an active role in eradication to disease surveillance.

In the mid-1950s, the following three events solidified the CDC's importance to U.S. citizens:

1. The CDC saved the polio program. In 1955, the clinical trials of Jonas Salk's (1914–95) polio vaccine had been

completed, and there was great public relief that this terrible *epidemic* might end. A nationwide vaccination program began, but soon children who had just received the newly approved vaccine became ill. The government halted the vaccination program, fearing that more children would become sick, and the CDC was called in to investigate. They traced the polio outbreak to one particular laboratory where the vaccine had become contaminated. With that discovery and the elimination of the problem, the inoculation program was resumed, thereby bringing to an end the polio epidemic in the United States.

2. The CDC began training disease investigators and establishing surveillance programs that are still vital today.

A photograph taken during a Hurricane Katrina planning meeting at the CDC in the Director's Emergency Operations Center (DEOC), August 29, 2005. The CDC's response to the disaster left by Hurricane Katrina involved working to support the affected states' public health response, under the leadership of the Federal Emergency Management Agency (FEMA). *(CDC)*

Langmuir began training groups of physicians who came to Atlanta to become disease detectives—a function still performed by the CDC. (See the sidebar "How the CDC Identifies an Illness" on page 7.) Among the surveillance programs they established was a continuous one for tracking influenza epidemics. Scientists began to see that the data they gathered could be predictive: By studying what viruses were floating around, they were able to predict the strains that they expected would predominate the following winter and in doing so prepare an appropriate vaccine. While the state of surveillance and prediction is quite good, it is not perfect. Most of the time the CDC does a relatively accurate assessment of what is anticipated, and each year the vaccine is modified to protect against the strain of flu that is expected to spread that year.

3. The CDC aided in the worldwide eradication of smallpox. The CDC also played a key role in the eradication of smallpox, one of the most notable achievements in public health. The CDC created and introduced to the World Health Organization a new way to vaccinate more people more quickly, and it has paid off. The last case of smallpox in the United States was in 1949 and the last naturally occurring case in the world was in Somalia in 1977, according to CDC statistics.

As medical professionals around the world have come to rely more heavily on this government entity, the CDC continues to become involved in more health issues worldwide.

Not a Perfect Record

Not everything at the CDC has gone perfectly. The CDC inherited the Tuskegee Syphilis Study in the 1970s and did nothing to bring that program to an end. A cure for syphilis had been found in the 1940s, so to maintain an ongoing study that involved not treating men with syphilis was mystifying and immoral. (See chapter 5.) In addition, the CDC predicted a huge outbreak of swine flu in

1976 and encouraged everyone to be vaccinated. Unfortunately, some people who received the vaccine developed a chronic illness known as Guillain-Barré syndrome. The CDC halted the vaccination program, and as it happened a different version of the flu appeared that winter—a swine flu *pandemic* did not occur. The growing concern in 2009 over whether or not the current H1N1 (swine) flu outbreak is going to become more virulent is complicated by what happened with the vaccination in 1976.

Today, the CDC also includes two particularly significant services—public education and worldwide disease surveillance—that were both outgrowths of programs that originated in separate Washington-based offices.

The public education role now filled by the CDC grew out of a program focused on venereal disease. When this Washington program was folded into the Atlanta offices, the CDC took on a specialty for which it is still known today: public education. The venereal disease program had featured employees who served as public health advisers, most of whom were not doctors but were trained to communicate health information. They became a mainstay of the fight against venereal disease by teaching people to guard against sexually transmitted illnesses. This role of educating the public on everything from healthy eating to the dangers of high blood pressure is a very important one played by the CDC today.

The worldwide disease surveillance program grew out of a Foreign Quarantine Service that was important in the early 20th century; the best way to guard against disease transmission at that time was to try to quarantine people before permitting them to enter the country in case they were ill. Today, this program has evolved to focus on overseas disease surveillance so that Americans will be aware of what illnesses may be traveling the globe.

In the 1970s, administrators wanted to address the expanded role of the CDC, and so they implemented a name change. At first, they referred to it as the Centers for Disease Control, then in 1992 the words *and Prevention* were added, though it is still often referred to as the CDC.

HOW THE CDC IDENTIFIES AN ILLNESS

In 1993, several perfectly healthy young people, notable athletes (distance runners), who lived in the Southwest on Indian reservations died quickly and unexpectedly. The disease was particularly virulent (those who became sick died of respiratory distress within 24 hours) and puzzling, so the CDC was quickly summoned to learn what they could and to keep it from spreading.

When the CDC representatives arrived on site, they knew they needed local input about what people were eating and doing differently that might have led to an outbreak in such healthy people of what was a terrifying disease. Because the deaths were on Indian reservations, the local people sent the CDC investigators to talk to the medicine men. The medicine men reported what they knew from their forefathers; this type of disease had occurred in their people three times in the last 100 years. The illness always occurred during years when there was an exceptional pine nut harvest. They also told the investigators that the bountiful supply of nuts was always accompanied by an exploding mouse population, which they then had to deal with. Each time these elements were present, there was also a bad type of human sickness.

At the same time, the investigators had sent tissue samples to a government laboratory where the disease was identified as a hantavirus, an acute illness that causes breathing difficulties and eventual kidney failure. Investigators knew that mice were the vector spreading the disease. The disease was originally identified during the Korean War when thousands of United Nations soldiers developed fever, headache, *hemorrhage,* and sudden kidney failure due to the exposure to field mice in Korea.

(continues)

(continued)

Hantavirus has now been identified in 25 states from New Mexico to Pennsylvania. It is one of the few viruses that can be transferred from rodents to humans through tiny droplets of urine or bits of freshly shed feces or in the dust and dirt that gets stirred up and then inhaled. It can also be transmitted from human to human.

The CDC has also provided the primary legwork in identifying the causes of two other particularly troubling illnesses of the 1970s and 1980s: Legionnaires' disease (a type of acute pneumonia that first occurred at an American Legion convention in Philadelphia in 1976) and toxic-shock syndrome (a rapidly developing, sometimes fatal disease that was identified in women using high-absorbency tampons during menstruation).

In the 1980s, the CDC also eventually unraveled information about *immune* deficiency disease, now identified as acquired immunodeficiency syndrome (AIDS). A significant portion of the CDC budget is still devoted to further understanding this disease.

For the last 30 years, the CDC has focused on applying scientific methods to the understanding of the spread of all types of disease. They focus not just on infectious illnesses, as they did at the beginning, but also on smoking, cancer, and the health effects of such things as hormone replacement therapy (HRT).

As early as 1950, government leaders were concerned about biological warfare. When the Korean War began in 1950, the CDC decided it needed an Epidemic Intelligence Service (EIS). Today, EIS is increasingly involved in how the government can effectively conduct bioterrorism surveillance.

The CDC is under the purview of the Department of Health and Human Services. The head of HHS is part of the presidential cabinet, and this governmental agency oversees food and drug safety, Medicare and Medicaid (see chapter 6), disease prevention, health information technology, financial assistance for low-income families to get health care, substance abuse and treatment, prevention of child abuse, domestic violence, and medical preparedness for natural disaster types of emergency and terrorism.

The U.S. surgeon general is also part of HHS, appointed by the president and confirmed by the Senate. While the job involves the supervision of a 6,000-person department that focuses on public health issues, most Americans see the surgeon general in his or her role as chief spokesperson on disease prevention or knowledgeable information on a particular outbreak anywhere in the world.

Another important aspect of government safety and health monitoring came about in 1970 when Congress passed the Occupational Safety and Health Act, creating the Occupational Safety and Health Administration (OSHA), a government body that is actually under the purview of the U.S. Department of Labor and provides an important view of work-related health and wellness in this country. Congress created it to establish and enforce

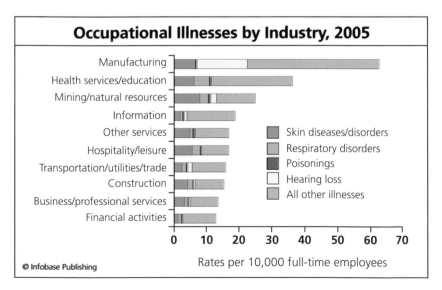

The workplace is one of the primary places where people are exposed to environmental and other health hazards.

workplace standards that help prevent work-related injuries, illnesses, and deaths.

The act that created OSHA also created the National Institute for Occupational Safety and Health (NIOSH), a research agency focusing on occupational health and safety. This agency is under the CDC. As defined by the CDC, NIOSH was established "to help assure safe and healthful working conditions for working men and women by providing research, information, education, and training in the field of occupational safety and health." NIOSH conducts all types of studies related to on-the-job accidents as well as the types of illnesses that seem to prevail in certain industries. As a result of these ongoing and continuous studies, the government receives accurate information on which to make decisions regarding legislation concerning the workplace.

MEDICAL FRONTIERS RECENTLY CROSSED

Medical success in the 21st century falls into many categories. As medical scientists have studied and perfected more and better ways to perform surgery, one of the main areas of focus has been cardiac surgery. From triple bypass surgery (where arteries or veins from elsewhere in the body are grafted to bypass a blocked coronary artery to improve blood flow to the heart) to using stents that are surgically placed to hold open blocked arteries, the changes in the field have been vast. Pharmaceutical discoveries have also improved heart health; an increasing number of Americans are now on statins, drugs that help keep cholesterol levels low. It is estimated that statins have reduced the risk of heart disease by 30 percent. Another heart advance has been technological. Inventors have successfully created a very simple automated external defibrillator (AED). Many workplaces, airports, and train stations now have these units on hand, saving countless lives. While organ transplants are discussed later in the chapter, it is certainly notable that heart transplants are performed more successfully than ever before.

Specific illnesses have received a great deal of recent medical focus. As mentioned earlier, smallpox has been totally eradicated (the last case was 1977), and while AIDS still cannot be cured, pharmaceutical cocktails are keeping patients alive and providing them with a better quality of life than before. Cancer has a similar story. While very few cancers can be cured through early diagnosis, a growing number of cases can now be treated, and patients can expect a relatively comfortable quality of life. At the same time, the medical profession is finding more and better ways to diagnose the various forms of cancer, and scientists are developing more targeted treatments that not only improve health but reduce side effects.

People are living longer lives, forcing medical science to address the needs of the aging. The ability to replace joints that wear out or are damaged is a medical art that has been perfected. Many of those who live into their 70s and beyond are undergoing knee and hip replacement surgeries. These surgeries return the patients to the ability to move around without pain. In addition, scientists are working hard to create better hearing aids and attempting to find remedies for vision issues such as age-related macular degeneration. These improvements will be vital as the population ages.

In addition, those who work in mental health care have made great strides in both diagnosis and medications. Psychiatrists now have many more and better options for mental illness than electroshock therapy. Today, doctors have many choices of *psychotropic* drugs that generally help how a patient feels or how he or she perceives the world.

Medical professionals have also encouraged greater public awareness of a multitude of issues that improve human health over time. With science behind them, the government and various nonprofit organizations have introduced helpful educational campaigns. The dangers of smoking have finally been better communicated to the public, and according to the American Cancer Society, the male death rate from smoking has dropped by 16 percent from 1991 to 2004. (These statistics are in the 2008 report, so they are the latest available.) Women were slower to give up the habit, so their statistical improvement is less dramatic. People now understand the importance of protecting themselves from the Sun. While diseases are still transmitted sexually, there is heightened awareness of the dangers posed by unprotected sex. An educational emphasis on the dangers of head injuries has resulted in laws in many states requiring the use of bike helmets for children, and this has reduced the rate of serious brain injury in bike accidents by 85 percent, according to the National Highway Safety Administration. Obesity is a growing problem (discussed in chapter 7) of the 21st century, and, if the public continues to ignore this, the rates of heart disease, diabetes, and many other illnesses will continue to increase.

TRANSPLANTS: A RECENT ACCOMPLISHMENT

The modern age of transplants began in the early 1900s, but at first there were very few successes. In 1905, Eduard Zirm (1863–1944), an Austrian physician, devised corneal transplants using corneas from cadavers, in the first successful transfer of an actively functioning body part. Zirm's success was with a farm laborer whose sight had been destroyed by exposure to a chemical. Zirm was aware that an 11-year-old boy had just died after an accident, so he retrieved the boy's corneas and transferred them to the farmworker. Zirm succeeded with only one of the two corneas transplanted, but the worker was able to see well enough to go back to work doing light farm duties. The method he created at the time forms the basis of the procedure used today.

In the late 19th century, physicians began trying to transplant skin. People were frequently hurt in fire-related accidents, and physicians tried to replace badly burned skin with patches from cadavers. In 1901, Karl Landsteiner had discovered that not all blood is the same; that blood must be categorized by type, so physicians worked to match blood types when *grafting* skin, but the host body frequently rejected the tissue. Doctors noted that if a second skin graft from the same donor was attempted, that graft was rejected even more quickly. Based on this finding, physicians

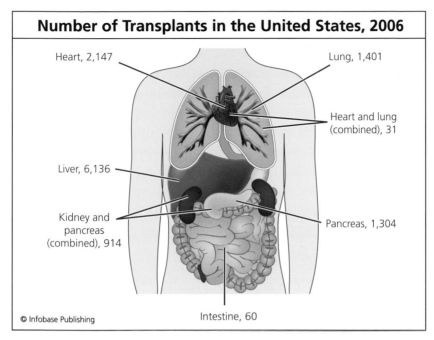

Number of Transplants in the United States, 2006

Heart, 2,147

Lung, 1,401

Heart and lung (combined), 31

Liver, 6,136

Kidney and pancreas (combined), 914

Pancreas, 1,304

© Infobase Publishing

Intestine, 60

Though success at performing transplants did not really take off until the 1980s when antirejection medications improved, this graphic displays the number and type of transplants that took place in 2006 in the United States. The transplant procedure has greatly increased in a relatively short period of time.

were beginning to understand that the patient's body built up an immunity if similar tissue had been previously introduced.

When Dr. Peter Medawar (1915–87) and Dr. Rupert Billingham (1921–2002) experimented with animals and skin grafting, they determined that if mice were inoculated at birth or even *in utero,* they could acclimate foreign tissue, and could tolerate grafting when they grew up. This provided proof that rejection of foreign organs could be overcome. Peter Medawar went on to win the Nobel Prize in physiology or medicine in 1960.

Kidney transplants were also explored. Because kidney disease is common and because kidneys come in pairs, it was actually the perfect organ on which to experiment. Kidneys can also survive well if kept cool until the organ can be put in place. (Patients with kidney failure cannot control their bodily fluids, and this causes an imbalance, which can be deadly because the accumulation of wastes can lead to a coma.) In 1914, scientists created a way to rid the body of the waste products the urine is supposed to get rid of, a process known as dialysis. Even today, dialysis is a lengthy and burdensome process, so the general hope was that dialysis could be used to help a patient long enough until a kidney transplant was performed. However, in 1914 no successful human kidney transplants were happening.

Physicians experimented with animals to try to master the process, and, although a few of the animals lived for a few days, none were cured successfully. The first human-to-human kidney transplant occurred in 1933 before the rejection issue had been solved. Then in 1947 a new challenge presented itself. A young woman had a uterine infection that resulted in her going into septic shock, and, as a result, her kidneys shut down. For 10 days she remained alive in a hospital in the Boston area but did not produce any urine. Finally, she was near death, so her doctor called on Dr. Charles Hufnagel, who had been experimenting with kidney transplants in animals. They planned to retrieve a kidney from a patient who had died, but when the doctors asked the hospital for the use of the operating room, the hospital expressed its disapproval of the procedure by turning them down. But Hufnagel's team had already obtained permission from the next of kin for the use of the dead patient's kidney, so they used a procedure room to perform the surgery. The doctors were aware that the kidney would be rejected, but they hoped the transplanted kidney would buy them time. After the transplant, the woman awoke and began to build strength and do better as the new kidney began to produce urine. A few days later, her body began to reject the new kidney but, amazingly, the process had jump-started her own system and her own kidneys were working again.

In 1951, at the same Boston hospital, Peter Bent Brigham Hospital, now the Brigham and Women's Hospital, physicians started having some success. One patient survived for five weeks after a transplant, and later that year one fellow whose transplanted kidney was placed in a plastic bag with the hope that this would prevent rejection was actually discharged and lived for six months. (Scientists later analyzed this situation and determined that the plastic bag had nothing to do with the success; the donor and patient had simply been a good match.) The pioneering surgeon who led the way on this was Dr. David Hume, who died in a plane crash in 1973 or would have shared in the Nobel Prize that was given in 1990 for successful organ transplant.

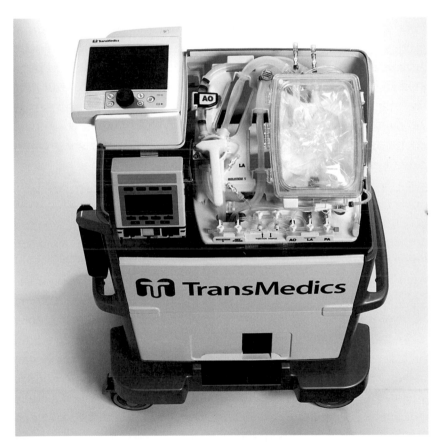

This is an example of the latest type of device for organ transport. It is designed to maintain human organs in a functioning state outside the body. Organs are kept in their physiological, functioning state during transport to the recipient and until implantation. *(The Heart and Diabetes Center North Rhine-Westphalia, Bad Oeynhausen, Germany)*

In 1954, an interesting occasion presented itself. A patient with renal failure who had an identical twin was admitted to Brigham Hospital. Doctors knew that twin-to-twin skin grafts could be done, so they wondered about trying the kidney. If this challenge were to be undertaken, it would be the first time a living person was asked to be an organ donor—something that is almost taken for granted now. The physicians decided to assess their assumption that it would work by first performing a small skin graft from twin to twin. Since that was successful, they then consulted the brothers about transplanting a kidney and both men agreed. In early kidney transplants, the surgeons tended to place the transplanted kidney without removing the person's own kidneys, and this is what they did with the patient known as R. H. The initial transplant went relatively well, but the physicians became concerned about R. H.'s blood pressure, and some other aspects of his blood tests did not look good. The doctors decided to see if removing R. H.'s own kidneys would solve the problem. Miraculously, it must have seemed, R. H.'s condition improved. He married one of his nurses and lived for eight more years.

Physicians knew that in the long run they needed to find a way to suppress the recipient's immune response to reject the foreign organ if organ transplants were going to offer any long-term solutions. They tried everything from injections of graft material from the donor to full body radiation treatments, as well as bone marrow transplants. (This is now a treatment used for cancer.) This experimentation began to show success when small and carefully regulated doses of radiation were used, and *cortisone* was prescribed if further flare-ups occurred.

Over time, doctors became aware that there were different aspects of the blood that needed to be examined and matched in order to increase the odds of acceptance. Even then, drugs are required, and today physicians have a number of *immunosuppressive* drugs to keep the host body from rejecting the organ.

Liver transplantation is more difficult than kidney transplants. The first successful one occurred in 1967, and two years later 10 patients (out of 69 attempts worldwide) had successfully received transplanted livers. Heart transplants were pioneered by Christiaan Barnard in 1967. (See *Medicine Becomes a Science,* the previous volume in this series.)

Locating organs was a constant problem, and finally the United States created a way to make it easier to find and get organs. In 1968, the Uniform Anatomical Gift Act (UAGA) established a uniform donor card as a legal document for anyone 18 years of age or older to legally donate his or her organs upon death. In 1984, the National Organ Transplant Act (NOTA) established a nationwide computer registry operated by the United Network for Organ Sharing (UNOS). It also authorized financial support for organ procurement organizations (OPOs) and prohibits buying or selling of organs in the United States.

THE FIRST TEST TUBE BABY

In early 2009, the world was shocked by the birth of octuplets to Nadya Suleman, an unemployed single mother. At first, the excitement swirled around whether so many babies could be born safely (physicians actually thought there were only seven; number eight was a surprise), and then the focus shifted. The babies had been born via in vitro fertilization to a single woman with no financial resources who was then revealed to already have six children. The ethics of the fertility specialist Dr. Michael Kamrava were called into question. Professionals and the public wanted to know why this procedure was performed on a woman who was unmarried, already had six children, and had no visible means of support. As the story deepened, medical eyebrows were raised by additional information. Suleman said Kamrava implanted her with six *embryos* for each of her six pregnancies—an apparent violation of national guidelines that specify no more than two embryos for a healthy woman under 35. In her last pregnancy, two of the six embryos split to create eight babies.

While national attention continues to focus on this mother of 14, few remember how recently it was that the actual process of implanting embryos was in itself highly controversial. The first test tube baby, Louise Joy Brown, was born on July 25, 1978 after great struggle. In the early 1970s, infertile couples had few options other than adoption. Not much was known about helping women become pregnant.

Finally, two British physicians had a breakthrough. Robert Edwards (1925–   ) served in the British army and was 26 before he applied to a graduate program in *genetics* at Edinburgh University. He became increasingly interested in the development of the embryo. His initial studies involved working with mouse embryos. He then went on to study *immunology* and fertility. In 1962, he asked to experiment with human eggs. However, the director of the agency where he was working told him not to pursue it—it was wrong. Edwards continued at Cambridge University, working with the eggs of cows, sheep, and monkeys, but he rarely had access to human eggs. In a case of what turned out to be beginner's luck, he managed to fertilize (with his own sperm) one human egg, but it was not to be possible again for years.

In 1966, Edwards met Patrick Steptoe (1913–88) a gynecologist who worked in Oldham, England. Steptoe was a pioneer in his own right and had spent years creating a laparoscopic method for doing gynecological surgery. Edwards explained to Steptoe his need for mature eggs from a woman's ovary in order to continue his experiments, and Steptoe was willing. The drive between Cambridge and Oldham was 165 miles, but Edwards repeatedly made the journey for years, looking for an answer. By the end of 1968, Steptoe and Edwards successfully fertilized some human eggs and followed their development through several cell divisions. As word of their work leaked out, they were chastised by the Catholic Church. The archbishop of Liverpool declared that what they were doing was morally wrong, and others wrote of the frightening specter of how this information might be used, worrying about selective breeding, *eugenics,* and *cloning.*

Steptoe found ready volunteers among women who had had difficulty conceiving, and the physicians turned their attention to finding a way to encourage a woman's eggs to mature so that they were ready to be harvested. After the eggs were fertilized in the laboratory (the test tube) and reimplanted at the right time. Their first attempt was in 1970, but it was unsuccessful. The publicity that followed them continued to be negative. The British Medical Council, whom they had approached for funding, noted "serious doubts about the ethical aspects of the proposed investigations in humans." When Edwards came to the United States, he was denounced by a leading theologian. Despite all the negativity, the Oldham health authority permitted them to continue their work. Expenses were covered out of Edwards and Steptoe's own pockets since no one was willing to fund them. Years went by while they tried to create a system.

With processes such as artificial insemination, the egg and sperm are joined in a test tube that is more likely a petri dish. This photograph shows a scientist carefully adding to the petri dishes she is supervising in her laboratory. *(Morphosys)*

Steptoe was older than Edwards, and by 1977 he was ready to retire. Failure loomed, and both men were exceedingly disappointed. They decided to give it one last try, and as they analyzed their procedure they determined that perhaps the process of encouraging a woman's ovaries to produce multiple eggs was part of the problem. They decided to see if a more natural approach would work. One of Steptoe's patients was Lesley Brown, a 29-year-old woman who desperately wanted children, and agreed to be a test case. Steptoe removed one single egg from Brown's ovaries using laparoscopy, and he used the husband's sperm to fertilize it. Two days later, the embryo was far enough along to be implanted in Brown's uterus.

The embryo was accepted by the uterus, and Brown was pregnant! Though Brown had some minor problems with blood pressure that caused some concern, a baby was delivered on July 25, 1978. Press interest was high. Today the story has come full circle. The baby who was born from this process, Louise Brown, conceived naturally, and her baby was born in December 2006.

Medical science has come a long way since the late 1970s, and social acceptance of in vitro fertilization has undergone a big change. While the Catholic Church still takes a strong stand against it, more and more people are taking advantage of it. The process is still costly, emotionally draining, and risky. Today, about one in three in vitro fertilizations results in a successful pregnancy, and because the process is difficult and costly physicians try to heighten the odds by increasing the number of embryos implanted. This has led to an increase in multiple births, which lead to babies being born prematurely at low birth weight or with heart and lung disorders, neurological problems, and developmental delays.

ADVANCES IN SURGERY

Robotic surgery (the use of robots to perform surgery) began in the 1980s. Because of the use of robotic arms that can move carefully within the small spaces of the human body, minimally invasive surgery, remote surgery, and unmanned surgery have all become possible. Major advantages of robotic surgery are precision, miniaturization, smaller incisions, decreased blood loss, less pain, and quicker healing time.

■ Minimally invasive surgery. For the patient, the advantage is that miniature hands can perform the work that previously had to be done by a human who almost certainly had larger fingers and hands. The less internal disruption the better it is for the patient. Another benefit is that robots are not limited to two hands. One of the most frequently used devices has four arms that are manipulated by the surgeon: One arm controls a camera that magnifies the internal organs so the surgeon has a clear picture of what is happening, and the other three arms are used for the surgery itself.

■ Remote surgery. In September 2001, Dr. Michel Gagner, working from New York, used a robotic system to remove the gall bladder of a woman in Strasbourg, France.

■ Unmanned surgery. In May 2006 the first unmanned robotic surgery took place. Unmanned surgery is the term used to describe a robot-guided process (rather than human-directed). A good example of it would be the threading of a catheter through a winding blood vessel or the precise placement of a needle where a very specific tissue sample needed to be removed.

Many general surgical procedures are now being performed robotically, for example, pancreatectomy or sectioning out a piece of a liver. Heart surgery is also being done robotically. Because

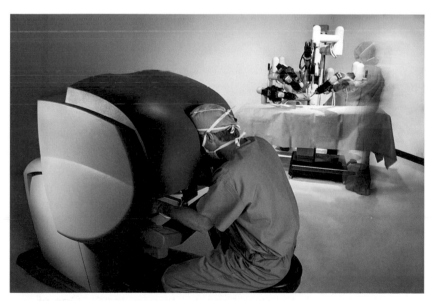

Here, a surgeon is using controls to manipulate a robotic device performing a surgery. The computer screen shows him what is happening. *(DaVinci, Inc.)*

the computer can provide a three-dimensional map of the heart and vasculature, it permits the robot to work carefully. The use of robots for minimally invasive gynecological surgery is also useful as it eliminates the need for large abdominal incisions. Neurosurgery, pediatric surgery, and surgery by urologists are a few of the other areas that are embracing robotic forms of surgery.

Further miniaturization of robots is actively underway. At Hebrew University in Jerusalem and the University of Nebraska Medical Center, scientists, engineers, and surgeons have been working collaboratively to create smaller and smaller robots that will minimize patient discomfort.

CONCLUSION

The CDC performs a vital function for the United States, with many governmental departments to monitor disease trends, conduct testing, be on the lookout for bioterrorism, and serve as part of a worldwide surveillance system. Parts of the CDC and the Office of Surgeon General are helpful in conveying information to the public. Job-related safety is also monitored and regulated by various governmental entities.

Some of the procedures that are almost taken for granted today, such as organ transplants, in vitro fertilization, and various types of advanced surgery are actually very recent developments. No doubt, there will be even more advances going forward.

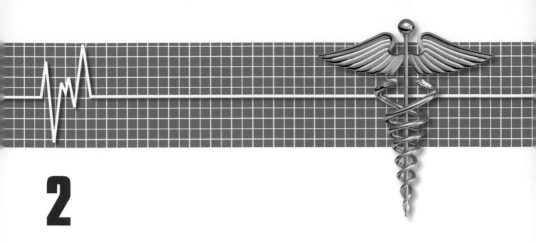

2

Human Health and the Environment

Most recommendations for good health habits focus on issues such as exercise and healthy eating. While these practices definitely contribute to better health, scientists are drawing attention to the fact that 25 percent of all preventable illnesses are caused by environmental factors. Proper management of the environment could take the world a long way forward in preventing disease.

People encounter damaging environmental problems in various ways. Both air and water pollution are major problems worldwide. According to a World Health Organization (WHO) assessment, more than 2 million premature deaths each year can be attributed to the effects of air pollution (both outdoor and indoor pollution). Air and water are only part of the problem; a good percentage begins with the way societies handle their garbage. For example, the practice of dumping garbage in landfills does not get rid of it. Certain types of waste in landfills become carcinogenic by-products that find their way into the air and water.

This chapter highlights some of the environmental pollution issues that are affecting human health. One of the most frightening situations has to do with mercury, a metallic element that

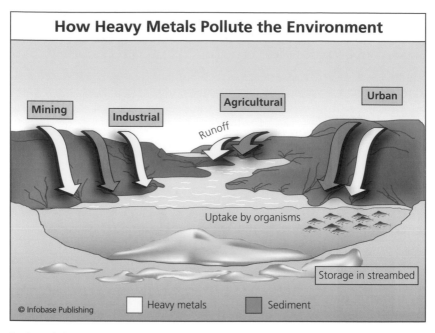

How Heavy Metals Pollute the Environment

Mining

Industrial

Agricultural

Urban

Runoff

Uptake by organisms

Storage in streambed

© Infobase Publishing

Heavy metals   Sediment

Industrial areas are not the only sources of metal and chemical air and water pollution. Metals and chemicals are released from all types of sources (trucks and cars on roadways are major sources of heavy metals; coal plants release enormous amounts of mercury). The pollutants become part of the air and drift to the ground where they land in urban as well as rural areas. Eventually, they mix with the runoff and enter major waterways.

played a prominent part in medical treatments of the past. Asbestos, lead, *radon,* and electromagnetic fields are discussed, as are the health issues that are predicted due to global warming. Ironically, medical waste is an issue that has received too little attention, and this too will be discussed.

THE DANGERS OF MERCURY

Today, mercury is known as an environmental *scourge,* but for centuries it was considered an effective medicine. Both Andrew Jackson and Abraham Lincoln took it regularly, and explorers Meriwether Lewis and William Clark carefully packed it into their first aid kit prior to leaving on their expedition to explore the Northwest. (See the sidebar "Mercury: Early Use as a Medicine" on page 28.) Mercury was also used as a dental amalgam for filling teeth, and many of today's adults have mercury-based fillings. The news that mercury is damaging to one's health is a relatively recent conclusion.

Mercury is a by-product of coal plants and some types of manufacturing. Once airborne, it settles into lakes and rivers where it accumulates in fish. Most people today showing dangerous levels of mercury have been exposed via fish consumption. Mercury settles into the body by adhering to fat, and because women carry 10 percent more body fat than men they are more prone to mercury poisoning. Children, too, are at great risk because their neurological systems are still developing.

Once in the human body, mercury acts as a neurotoxin, interfering with the brain and the nervous system. In 2005, the Centers for Disease Control and Prevention (CDC) estimated that one in 15 U.S. women of reproductive age have a blood mercury level above 5.8 micrograms per liter of blood—a level that could cause birth defects in a developing fetus. A California physician performed her own study and found that the mercury in her female patients' blood was actually 10 times higher than the CDC's average reading; in children the level was sometimes 40 percent higher.

Mercury in Fish

To date, 48 states have issued advisories on mercury in fish. In August 2004, the U.S. Environmental Protection Agency (EPA) announced that fish in virtually all U.S. lakes and rivers were contaminated. The U.S. Food and Drug Administration (FDA) advises that pregnant women, women who may become pregnant, nursing mothers, and young children avoid shark, swordfish, king mackerel, and tilefish entirely, and they should limit consumption of albacore tuna (canned white tuna and tuna steaks) to six ounces (one meal) per week. (These fish are at the top of the food chain. Because they consume smaller fish in mercury-polluted streams, these bigger fish contain higher levels of mercury.) Canned light tuna, shrimp, salmon, pollock, and catfish are said to be tolerable if a person eats no more than 12 ounces per week.

Scientists have also determined that the level of mercury pollution is much more widespread than originally thought. In early 2005, a study in the journal *Ecotoxicology* found high levels of mercury in songbirds, salamanders, and other New England wildlife previously thought to be unaffected. This was discouraging news because up until recently scientists thought that the mercury poisonings were limited to species that consumed the toxin directly from water. This new evidence indicates that the mercury emissions from sources such as power plants travel surprisingly long distances and drift down onto plants on land, where it is consumed by sow bugs, centipedes, and other small insects. These are then ingested by birds and other wildlife, thus spreading the toxicity of the pollution.

If high levels of mercury in the body are detected early, most people can bring their levels down by reducing the amount and the kind of fish they consume. In more severe cases, a chelation process can be helpful in ridding the body of this toxin. For people—if not for industry—the preferable long-term solution would involve reducing the levels of mercury in the environment.

The Harsh Lesson Learned in Japan

If the dangers of mercury were not fully understood early in the 20th century, they became abundantly clear in Japan during the 1950s. Minamata, a town located on the western coast of Japan's southernmost island, was downstream of the site of the Chisso Corporation, located in Kumamoto, which by the mid-1930s was manufacturing acetaldehyde, a substance used to produce plastics. From 1932 to 1968, Chisso dumped an estimated 27 tons of mercury compound into the bay, and it took decades for anyone to realize that the heavy metal had transformed into methylmercury, an organic form that easily enters the food chain.

Because the town was located near the water, Minamata residents consumed fish regularly as did stray cats that seemed to be able to obtain their own fish dinners. The cats were the first to raise public concern. The felines began to show signs of erratic

behavior, including dizziness and decreased motor skills, and people referred to them as dancing cats. These symptoms soon began
to display themselves in people, who also reported feeling poorly
and found that they often could not speak without slurring their
words. Ultimately, more than 3,000 residents showed the effects
of mercury poisoning; 46 people died.

The Different Forms and Uses of Mercury

Mercury is found both in its elemental state and in organic and
inorganic compounds, and it is present in the environment as a
result of human activity as well as from natural sources such as
volcanoes and forest fires. It has been used for 3,000 years both in
medicine and industry. There are three main forms of mercury.

1. Elemental mercury is the silvery liquid used in thermo-
   stats, barometers, batteries, folk medicine, and by experimenting students in chemistry class. It vaporizes quickly
   when heated and is toxic in its vaporized form.
2. Inorganic mercury salts are used both in antiseptic
   creams and ointments and in electrochemistry.
3. Organic mercury compounds such as dimethylmercury
   or methylmercury are created when mercury in the
   air lands in water—or on land and is washed into the
   water—where bacteria can change it into methylmercury,
   a highly toxic form of the substance that builds up in fish,
   shellfish, and animals that eat fish. These are by far the
   most dangerous forms of mercury as they can be readily
   absorbed by the body.

In the United States, the use of mercury caught on during the
gold rush when men found it could be used in the amalgamation of
gold and silver. Then, as the Industrial Revolution got under way,
people found all types of uses for mercury, including the creation
of daguerreotypes, the silvering of mirrors, and as a preservative.
As a matter of fact, because of its preservative qualities, it was a
valuable ingredient in wall paints up until 1978, when its use was
outlawed.

MERCURY: EARLY USE AS A MEDICINE

Mercury in the form of its natural reddish pigment, known as cinnabar, was used for cosmetic purposes by ancient Egyptians and Chinese, and the Greeks created medicinal uses for it, which were carried forward by other people. From the 16th century right up until the mid-20th century, mercury salts were a primary treatment for syphilis. Presidents Andrew Jackson and Abraham Lincoln are known to have taken mercury. Researchers analyzed preserved samples of Jackson's hair from two different time periods (1815 and 1839) and both showed high levels of mercury. Historians report that many of Jackson's physical problems—digestive ailments, excessive salivation, headaches, hand tremors, and dysentery—may have actually been caused by his mercury treatment.

Recent accounts have been written about Lincoln's bouts of depression, and historians talk of his use of the blue mass pill. This pill was a combination of licorice root, rose water, honey, sugar, and a confection of rose petals, but its main ingredient was elemental mercury in enough quantity that it could have caused mood swings, tremors, and neurological damage. (The typical blue mass pill is thought to have contained 9,000 times more mercury than is safe.) Fortunately for history, Lincoln stopped taking these blue pills shortly after his inauguration.

Early in the 20th century, mercury was given to children annually as a de-wormer, and it was sometimes used in teething powder for infants. It was also used in mercurochrome, a popular medicine that was dabbed on cuts and scrapes. Mercurochrome predated federal oversight on medications, and it took the government a long time to work its way back through medicines that were already being marketed. In 1982, the FDA began a review of mercurochrome, but it was not until 1998 that the FDA finally pronounced that it was "not generally recognized as safe and effective" as an over-the-counter antiseptic and prohibited its sale across state lines.

The discussion about the safety of filling teeth with a mercury-containing substance has been a long one. As early as 1843, the American Society of Dental Surgeons required members to sign a pledge that they would not use it; however, an amalgam containing mercury is still sometimes used today. Although some health activists claim the mercury leeches out of the fillings and into the body, an FDA 2002 press statement reaffirmed the mainstream view: "No valid scientific evidence has shown that amalgams cause harm to patients with dental restorations, except in the rare case of allergy." Ironically, in 1988 scrap dental amalgam—fillings that have been removed from the mouth—was declared a hazardous waste product. In an attempt to lead the nation in a more positive direction, California became the first state to attempt to ban the use of mercury fillings.

The most recent issue concerning mercury and medicine has been its use as a preservative, known as thimerosal, in childhood vaccines. At the same time that the number of required vaccines has increased, there has been an increase in the diagnosis of autism in children. Though studies have not proven a link between thimerosal and autism, scientists do not yet know why there are more cases of this developmental disorder. There is agreement that mercury is a neurotoxin that can cause great harm to a developing nervous system. As a result, the National Resources Defense Council (NRDC) and others successfully pressed for the removal of thimerosal from childhood vaccines. Now all vaccines are available in a mercury-free form, though some flu vaccines still contain it as a preservative, and some forms of the diphtheria and tetanus vaccines have trace residues.

Mercury is also used in the manufacture of industrial chemicals or for electronic and electrical applications. Gaseous mercury has been used in some forms of neon sign advertising, and liquid mercury is sometimes used as a coolant for nuclear reactors.

The Sources of Mercury Emissions Today

Coal-fired power plants are the largest single source of mercury emissions in the United States, accounting for more than 90,000 pounds of airborne mercury a year—about one-third of the total output. Manufacturing plants that produce chlorine formerly used a mercury-based process that released a great deal of mercury pollution. However, a 21st-century environmental campaign has encouraged many of U.S. chlorine plants to adopt mercury-free processes, though there are still a good number of plants in Europe that use the mercury-based method.

Until 2001, factory and power plant emissions were governed by the Clean Air Act, which required plants to have the best available technology in place by 2009. The improvements were projected to lower emissions by 90 percent. The Bush administration changed course, removing the power plants from Clean Air Act jurisdiction and proposing the first regulatory effort to cut the emissions with a plan to reduce output by 70 percent within 13 years. In addition, the EPA also allows the power companies to buy pollution credits instead of reducing emission levels. In 2008, the U.S. Court of Appeals for the District of Columbia issued an opinion in a case that was initiated by 15 states and other groups, challenging the EPA's decision to delist mercury as a hazardous air pollutant. While this—and the arrival of the Obama administration—makes it likely that a more stringent mercury standard will be adopted, EPA rulemaking takes time.

LEAD AND ASBESTOS

Lead and asbestos are used for very different purposes, but they have two elements in common: they were viewed as extremely useful by early civilizations and are now recognized as very hazardous to human health. Both lead and asbestos were used as early as Greek and Roman times.

Lead

Lead is a natural substance that is relatively easy to extract from ore, and, once extracted, its low-melting temperature makes it easy to work with. It does not rust and is very durable, so its use grew quickly. The Romans found that lead was an ideal substance for their extensive piping system (the word *plumbing* comes from the Latin word for lead *plumbum*), because it was easily available and very malleable when heated. Lead was so popular for creating things that it soon became the basic ingredient for coins and pewter dishes and other household artifacts. It was added to face powders, rouges, and mascaras, and chefs for the wealthy used it to halt the fermentation of wine and even used it as a food seasoning.

Though it did not halt their use of it, ancient people noted that those who worked with lead regularly became crazy; they referred to it as lead intoxication. Some modern scholars believe that lead poisoning contributed to the fall of the Roman Empire. Miners and smelters suffered ill effects from lead exposure, but the wealthy were affected since they were more likely to be exposed to the implements created using lead as well as its use in food preparation.

Mining and smelting of lead was introduced early in the American colonies, and by the 20th century the United States was producing and using more lead than anywhere else in the world. Though countries such as Australia were realizing some of the hazards of lead and were beginning to ban it for certain uses, inventors in Europe and America were looking for even more ways to put lead to use.

In the 1920s, inventors thought of adding lead to gasoline; it boosted the octane levels and reduced the sound of engine-knocking. Unfortunately, as cars burned the gasoline, lead by-products polluted the air. The full impact of the problem occurred at several refineries in the United States during this period. In the summer of 1924, workers at more than one U.S. location where the additive was produced became sick; a total of 15 people died. The following

May (1925), the U.S. surgeon general suspended the sale of leaded gasoline and created a panel to study the issue. The committee primarily consisted of industry executives, with one exception: Dr. Alice Hamilton of Harvard (Hamilton is discussed at length in *Old World and New,* another book in this History of Medicine set) whose specialty was work-related health hazards. The committee was given only seven months to conduct their study. In 1926, the committee issued a report indicating that given the length of time they had studied the issue, they did not find sufficient evidence to ban the use of lead as an additive. The authors of the report included a qualifying paragraph noting that if the widespread use of gasoline were studied over a longer period of time, then evidence of the dangers of lead exposure to humans might be revealed. As it was, not enough evidence was found, so in 1927, the surgeon general again permitted lead-based gasoline to be made. However, a voluntary limit of lead content was recommended.

By the 1960s, the authority over lead emissions fell under the Clean Air Act and eventually the newly created EPA (1970). Its first administrator, William D. Ruckelshaus, noted the need to reduce lead content because of "an extensive body of evidence" that showed it was a threat to human health.

Lead was banned as a gas additive in the 1980s and has not been used in paint in the United States since the 1970s, but elevated lead levels in people remain a problem. Lead continues to be used in many products, including batteries, ammunition, solder, pipes, pottery glazes, printing inks, and paint for industrial, military, and marine use. Lead used in the past does not break down, so lead has contaminated soil in various parts of the country, particularly near major roadways where lead was released in car exhaust and in older neighborhoods where lead paint was used for so long.

While scientists now know that lead exposure is not healthy for anyone, they have noted that children's exposure seems more damaging because their bodies are smaller. Children's greatest exposure is from paint chips. If they live in an old house where the paint is chipping, they may ingest the chips or a crawling baby may pick up traces on his or her hands from dust on the floor. These children frequently show signs of learning delays or disabilities. Some recent reports have shown small declines in IQ scores in children who have had early exposure to lead sources.

Universal screening for lead levels in people is not required, but the CDC recommends testing children who live in older neighborhoods, children who have lived in a home under renovation, or children with symptoms. Sometimes medication must be used to reduce lead levels, but in general lead levels can be reduced by removing the source of the lead exposure and adding foods that are high in calcium and iron into the diet. These substances help keep lead from being absorbed in the body.

Asbestos

Just as the serious dangers of mercury and lead were underplayed for many years, a similar lack of attention surrounded the use of asbestos. Asbestos is a naturally occurring material made up of microscopic bundles of fibers that can become airborne when disturbed. When these fibers are released into the air, they may be inhaled into the lungs, where they are capable of causing significant health problems.

The first recorded use of asbestos was by the Greeks who valued its fire retardant properties, but they also noted its ill effect on the lungs of slaves who were assigned to weave it into cloth. During the Middle Ages, asbestos fell out of favor, but it became popular again during the Industrial Revolution when it was used to insulate steam pipes, boilers, kilns, and ovens. The negative effect on human health was either ignored or forgotten.

According to the Asbestos Resource Center, the first time researchers began to note that asbestos workers were suffering common—and often deadly—health problems was in 1917–18, when men in mining communities began dying of lung problems. In England in 1924, a 33-year-old British woman who had worked with asbestos since she was 13 died of a lung ailment that her doctor identified as asbestosis, a condition where the asbestos fibers become trapped in the lungs and make it difficult to breathe. As a result of this diagnosis, the British government undertook a study

The damaged insulation on the pipe exposes workers to loose asbestos fibers. *(Bath and Northeast Somerset Asbestos Registry, U.K.)*

of asbestos workers. In 1931, Britain deemed that asbestosis was a legitimate work-related disease and passed a law that required asbestos plants to have better ventilation to try to reduce human exposure to the fibers.

In the United States, manufacturing companies took another 10 years to implement similar rules in industries dealing with asbestos. Though the dangers of working with asbestos were recognized, no one focused on the fact that it also might be hazardous to people whose exposure was environmental and not work-related. Over time, scientists began to see that in locations where the asbestos became damaged or disturbed in some way, it could be harmful to people's health. In what was a rather slow response to this news, there was some reduction in the dependence on asbestos during the 1970s, but it was not until 1989 that laws were put in place to restrict the use of asbestos in building materials. Then, in 1991, the laws were weakened. The Fifth Circuit Court of Appeals in New

Orleans overturned the EPA ruling that more severely limited the use of asbestos. The court left in place the banning of flooring felt, rollboard (a product used to create ceilings and walls in primarily commercial construction), and corrugated, commercial, or specialty paper, and banned the use of asbestos in any new uses.

Today, most schools, office buildings, and homes that are undergoing some level of renovation must first hire a company to

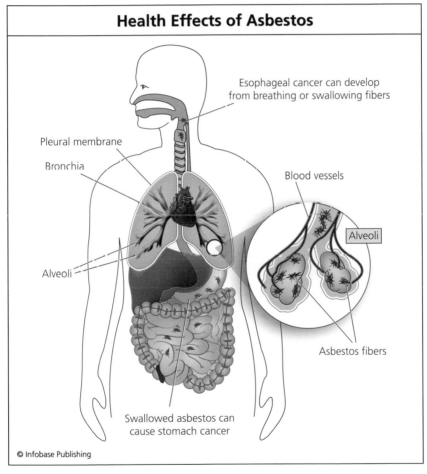

Health Effects of Asbestos

Esophageal cancer can develop from breathing or swallowing fibers

Pleural membrane

Bronchia

Blood vessels

Alveoli

Alveoli

Asbestos fibers

Swallowed asbestos can cause stomach cancer

© Infobase Publishing

If the asbestos becomes damaged, then the microscopic fibers can become airborne. Physicians compare asbestos exposure to exposure to something like cigarettes—the more the exposure, the greater the likelihood of health problems.

perform asbestos abatement, which involves creating a sealed tent-like workspace and workers dressed in spacesuit-type gear to go in and remove the asbestos in the area where work is to be done.

Medical professionals note that, just as in cigarette smoking, the greater the exposure, the greater the health hazard. Certainly those who work with asbestos are placed at greatest risk, followed by anyone living or working in an area where the asbestos is not encapsulated by other materials. In addition to asbestosis, people can have mesothelioma (a cancer of the outside tissues of the lungs that is caused by asbestos) or lung cancer as a result of exposure. From the time of exposure to the development of an illness can take anywhere from 15 to 40 years. Those asbestos workers who also smoke are at the greatest risk of getting lung cancer.

RADON AND ELECTROMAGNETIC FIELDS

These two elements are currently at opposite ends of the health spectrum. Radon, a naturally occurring radioactive gas, has actually been determined to be quite damaging to human health. Electromagnetic fields, a relatively recent phenomena about which there has been great concern, are currently considered not hazardous.

Radon

Radon is a cancer-causing radioactive gas that exists in various parts of the United States. It enters the atmosphere from the natural breakdown of uranium in soil, rock, and water. While radon has always been around, it became recognized as a carcinogen as people began living in relatively airtight housing. The radon gases become trapped within the structure and are particularly harmful as people breathe in accumulated levels of the gas.

The U.S. surgeon general warns that radon is a leading cause of lung cancer in the United States today, second only to smoking. Radon is responsible for about 21,000 lung cancer deaths every year, about 2,900 among people who have never smoked. The EPA encourages all home dwellers to test for radon as living with it in the home is particularly dangerous. There are very simple home radon test kits on the market, and if a high level of radon is found, there are ways for the problem to be remedied.

Electromagnetic Fields

As the modern world becomes increasingly electrified, population centers are surrounded by power lines, and giant cell phone towers beam radio waves. Within homes, people are surrounded by computers, television sets, appliances, and electrical wiring, all of which emit electromagnetic fields (EMF) at various levels. Some people have feared that everything from cataracts to cancer may come from this low-level radiation.

The initial alarm over electromagnetic fields began in 1979 when a study was conducted that indicated that childhood leukemia was higher in communities with higher EMF fields. Since 1979 and the original study, many scientists have conducted studies to evaluate the findings, and current reports point out that

Up until about a decade ago, there was a great fear that power lines were hazardous to human health. This photograph shows power lines in New England. *(Wikimedia)*

the original study had major flaws. More recently, organizations like the National Institute of Environmental Health Sciences (NIEHS) (part of the National Institutes of Health [NIH] of the U.S. Department of Health and Human Services [HHS]) report that electric and magnetic fields should be regarded as "possible human carcinogens," but that there is no reliable evidence to draw the same conclusions that the original study did. As a matter of fact, the incidence of childhood leukemia has recently been dropping, and this is counter to what is happening in society with an ever-increasing number of ways that people are using power and electricity.

GLOBAL WARMING EXPECTED TO AFFECT HUMAN HEALTH

In addition to all the environmental issues affected by global warming, scientists warn that disease patterns will inevitably change as average temperatures of the world continue to increase.

"Environmental changes have always been associated with the appearance of new diseases or the arrival of old diseases in new places. With more changes, we can expect more surprises," Stephen Morse of Columbia University told the 107th general meeting of the American Society for Microbiology in Toronto in May 2007, according to a press report issued by the organization. Examples of the expected types of changes include:

- The pattern of insect- and tick-borne diseases such as malaria, yellow fever, and Lyme disease will change. Insects are highly sensitive to alterations in vegetation, temperature, and humidity levels, so as the average temperatures increase, the course of transmission of these types of diseases will inevitably change.
- Malaria, already a terrible scourge in developing countries, will be even more difficult to stop. With the current average climate temperatures, malaria is not transmitted

in higher altitudes because of cooler temperatures. As temperatures rise, malaria will spread within a wider area.

■ In the Tropics, influenza exists year-round, not seasonally the way it does in climates with more variable weather. As the seasons moderate around the world, this may mean that flu season is continuous. The 2009 outbreak of H1N1 flu defied seasonal norms with many people coming down with the flu during the summer months. Whether or not this is a continuing trend that was set off by global warming will have to be analyzed later.

■ Extreme weather has a major effect on the infrastructure in all countries, but in developed countries where the population is accustomed to easy access to clean water and disease-free streets, the disease patterns alter drastically when infrastructure is damaged. The havoc wreaked on New Orleans by Hurricane Katrina in 2005 offers an example of what can happen. When the areas flooded, sludge and sewage were carried into the reservoirs, making the water supply unusable. Sewage in the streets meant that disease could spread as people moved around the city. In addition, people who have to evacuate their homes are then housed in schools or community centers for the duration of the emergency, and this contributes to the spread of disease.

■ As the ozone layer is depleted through global warming, people on earth are exposed to higher levels of this oxygen-related gas, which at ground level is a harmful air pollutant and a main contributor to urban smog. Long-term exposure can cause a reduction in lung function, inflammation of the airways (see the sidebar "Rise in Asthma Partially Due to Pollutants" on page 40), and respiratory problems.

■ Indirect changes to human health may result from changes in agricultural practices. Drought or high temperatures may change the ability to produce food, and this, in turn, will affect people's ability to maintain good health.

RISE IN ASTHMA PARTIALLY DUE TO POLLUTANTS

Asthma is a common breathing disorder in which inflamed airways make it hard for a person to breathe, causing wheezing, shortness of breath, and coughing. This illness is on the increase. Over the last 10 years, researchers report a substantial rise in the prevalence and severity of asthma in children, and scientists are documenting more cases of asthma among children in areas where air pollution is high, leading experts to believe that there is a direct cause between asthma and dirty air. An NIEHS-funded study shows that children who played three or more outdoor sports in areas with high ozone concentrations (a form of pollution) are three times as likely to develop asthma as other children. However, the National Heart, Lung, and Blood Institute (NHLBI), another part of

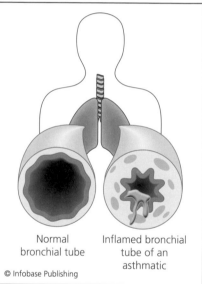

Why Asthma Makes It Hard to Breathe

Normal bronchial tube

Inflamed bronchial tube of an asthmatic

© Infobase Publishing

When the bronchial tubes become inflamed because of an irritant, their air passageway is reduced, making it more difficult to breathe. If an asthma attack is not managed properly, it can result in a trip to the emergency room.

While it will take time to assess how much environmental changes lead to new health problems or worsen illnesses that already exist, there is no doubt that change is underway.

the NIH and the HHS, stresses that pollution alone does not cause asthma. Asthma usually is the result of a combination of factors:

- Family tendency to have allergies or asthma
- Certain respiratory illnesses during childhood can predispose a person to have asthma
- Exposure to airborne allergens or pollutants

All of these factors can make a person's airways more reactive to substances in the air and result in asthma.

Medications that keep the bronchial tubes dilated are helpful. Those who are diagnosed with asthma are prescribed inhalers for faster relief; symptoms are milder when the illness is dealt with right away. Countless doctors tell of middle-of-the-night emergency room visits for children who could not breathe—unnecessary emergencies if medication had been started at the first sign of recurrence of the disease. Though the illness generally starts when people are young, those with asthma must manage the illness for the rest of their lives.

Inhaler in use *(Hideki)*

MEDICAL WASTE

Though the medical community has every right to expect other industries to help reduce environmentally damaging waste, the medical industry itself needs to develop a worldwide plan for dealing with health-related waste products.

With a surprising level of regularity, medical waste is found on American beaches, and that is only the type of medical waste that is visible. People today are exposed to medical contaminants through many other sources. Medical material that goes into a landfill may result in contaminated drinking water. Those that are incinerated are released into the air. (When anything containing chlorine is burned, it turns into a carcinogen.) Anything with heavy metal content—lead, mercury, and cadmium—results in the spread of heavy metals in the environment. Radiation is widely used in various aspects of medicine, and disposal of this potentially very dangerous waste is handled in many different ways. In addition, pharmaceutical products, particularly *antibiotics* and *cytotoxic* drugs, can become part of a waste stream and add to the pollution in soil and water.

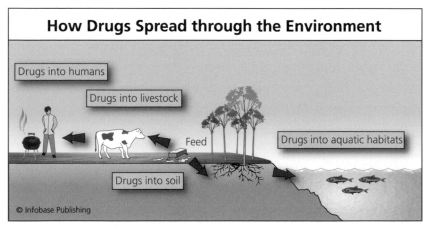

How Drugs Spread through the Environment

Drugs into humans

Drugs into livestock

Feed

Drugs into aquatic habitats

Drugs into soil

© Infobase Publishing

Just as any other waste product mixes into the environment so too do pharmaceuticals. The drugs given to livestock as well as drugs for humans (particularly unused medicines that people commonly flush down toilets) mix into the environment and reappear in water sources.

The risks involved in health care wastes have not been well studied, but it is estimated that approximately 20 percent of medical waste involves hazardous materials that may be infectious, toxic, or radioactive, according to information produced by WHO. Infectious waste (ranging from contaminated human waste to laboratory materials used on highly infectious individuals and anatomic animal or human parts) constitutes about 15 percent of hazardous waste, expired or contaminated pharmaceuticals and chemicals (solvents and disinfectants) make up about 3 percent of the hazardous materials, and the remainder consists of syringes and other sharp items, genotoxic waste—carcinogenic materials or certain drugs used in cancer treatment and radioactive matter—and waste with heavy metals such as mercury.

An example of what can happen with medical waste occurred in Vladivostok (Russia) in 2000. WHO reports that six children found glass *ampules* containing some expired smallpox vaccine in an area where they were playing. All six came down with a mild form of the disease. While none suffered illnesses that were life-threatening, it certainly points to the need for careful disposal.

Currently there are few environmentally friendly and affordable methods for disposing of infectious wastes, according to WHO. Developed countries have progressed to the point that they separate hazardous from nonhazardous waste; poorer countries do not. Different bio-products require different disposal methods: Some should be chemically treated, others can be destroyed by burning at a very high temperature, still others require autoclaving, which complicates matters.

To remedy the situation, scientists who have been studying this problem say this needs to be a worldwide effort. They recommend that countries work together to determine appropriate rules for handling medical waste. Guidelines should be written that specify how best to dispose of various types of waste and in what way people undertaking this disposal should dress to protect themselves. This information emphasizes the importance of WHO's goals to encourage each country to develop health care waste management methods and to create a database with recommendations of the best way to dispose of various waste materials.

The Evidence in the Water

A CBS News report on an Associated Press (AP) investigation (March 10, 2008) revealed the presence of a wide variety of pharmaceuticals—antibiotics, anticonvulsants, mood stabilizers, and hormones—in American drinking water. While the amounts of any single drug in the water supply are minute, scientists are concerned because of the high number of prescription and over-the-counter drugs that have been found.

Most of the drugs find their way into the water supply because people tend to flush medicines they do not take into the toilet. While all wastewater is processed at a water treatment facility, current treatment methods do not remove all of the drug residue. Also, treatment facilities—while required to check for a wide variety of metals, pathogens, and various by-products—are not required to test for miscellaneous drugs. Yet, the AP investigation revealed the presence of the following in water samples taken in 2008:

- In Philadelphia, 56 pharmaceuticals or by-products were found in the water supply, including pain medication, antibiotics, cholesterol-lowering drugs, epilepsy, and asthma medications. In the *watershed* areas around the city, 63 pharmaceutical by-products were found. (Drugs can enter the watershed through leaking septic systems or through medications given to animals.)
- In Washington, D.C., six types of pharmaceutical products were found in the water.
- Water delivered to 18.5 million people in southern California had both antiepileptic and antianxiety medications detectable.
- In Tucson, Arizona, an antibiotic and two other medications were clearly found in the water.

The interior of Robinson Helicopter Factory presents a clean workplace that is likely free of environmental hazards. *(Robinson Helicopter Company)*

When interviewed on this subject by CBS News, Dr. David Carpenter, director of the Institute for Health and the Environment at the State University of New York at Albany, reminded people that bottled water is no better. Many bottling companies simply bottle tap water. Carpenter notes that a consumer's best plan is the charcoal-filtering systems that can be used at home. However, consumers need to follow directions carefully and change the filters frequently, or they may create another type of health problem.

CONCLUSION

In the Middle Ages, towns sometimes dumped their waste into one part of a river and then drew what they thought was freshwater from a different part of the waterway. Over time, of course, they polluted their own water sources. While scientists today are much more knowledgeable about pollution than were the people of the Middle Ages, there arc still similarities in how communities live their lives: Coal plants produce mercury, and mercury contaminates fish eaten by humans who then suffer various levels of mercury poisoning. Yet despite this negative cause and effect, it has been very difficult to bring about change in the world.

While certain environmental pollutants are somewhat easier to monitor and regulate, there are many ways society continues to damage the world. Asthma and cancer are just two of many diseases where the causes can partially or totally be traced back to the environment. There is no doubt that as the world looks for ways to improve human health, leaders will need to pay attention to the environmental issues that affect human well-being.

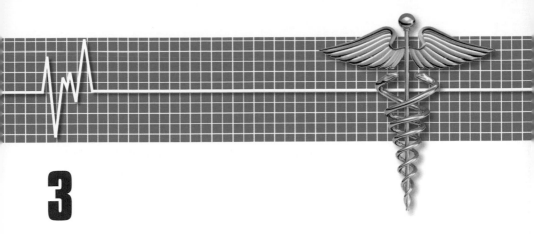

3

Animal and Human Diseases— a Definite Link

In the last 30 years, the number of new diseases that have jumped from wildlife to humans has increased, and many of these illnesses have proven to be quite harmful to humans. HIV/AIDS, Ebola, Lyme disease, avian flu, swine flu, West Nile virus, monkeypox, and SARS are among them, and scientists are reckoning that future study must closely track animal diseases to be prepared for the possibility that these could transfer to humans.

Because of rapidly changing human behavior and animal ecology, infections are spreading more widely and more quickly than before. The difference in the climate for illness seems to rest in the more dense groupings of animal species, an environment that makes it possible for pathogens to jump among animal types. This environment is unique to modern life. When countries were more sparsely populated, there was ample room for both people and animals, so diseases did not spread among species. Today, scientists report that more than half of all emerging diseases have started in the animal population, and these illnesses have eventually been transmitted to people. These diseases emerge because of changes in human activities, such as wildlife trade and global travel.

This chapter highlights some of the more recent and deadly illnesses that have started in an animal population and then spread to have devastating effects on humans. SARS (severe acute respiratory syndrome) and HIV/AIDS (acquired immunodeficiency syndrome) will be touched on, and a more complete look will be given to the continuing concerns about avian influenza as well as the growing concern over the recent occurrence of the strain of influenza A known as H1N1 (or the swine) flu pandemic. While the disease has been in the news less in the last year or two, it has not gone away, and the scientific approach to disease surveillance and the attempts to arrest the illness if it does move into the human population in a major way are very instructive.

SARS—A CLEAR ALERT

The SARS outbreak of 2003 offers a perfect example of a 21st-century disease that knows no borders. While in earlier times disease outbreaks might have remained local, today an outbreak in one country can very rapidly become a problem for countries on the other side of the world.

SARS poster *(The Chinese University of Hong Kong)*

In 2003, SARS dominated the headlines of every major news organization. The disease quickly spread to 30 countries in nearly as many days, claiming more than 800 lives and terrifying millions of people. People all over the world were spooked by the news coverage of the citizens of Hong Kong and the major cities in eastern China who wore face masks everywhere to try to prevent transmission of the illness. Travel restric-

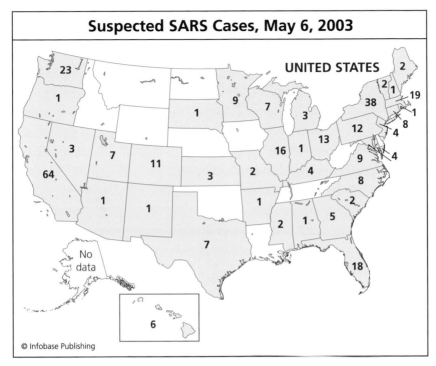

Suspected SARS Cases, May 6, 2003

UNITED STATES

23
1
1
9
7
3
2
2
1
38
19
1
8
4
3
7
11
16
1
13
12
9
4
64
3
2
4
8
1
1
1
2
2
1
5
7
No data
18
6

© Infobase Publishing

From November 2002 to July 2003, health officials were very concerned about the spread of SARS, so they were carefully tracking its occurrence.

tions were put in place, and the disease crippled tourism in several countries and caused many other ripple effects from school closures to quarantines to various forms of discrimination based on fear that certain people might be harboring the illness. Fear gripped the United States when an outbreak occurred in Toronto, proving that the illness could move out of Asia and go almost anywhere.

The slowing of SARS began when scientists traced the source to a coronavirus that was harbored by civets and raccoon dogs. From there, scientists identified bats as the ultimate source, but they felt that animals confined in Chinese marketplaces may have been the way it spread to people. Marketplaces that featured wildlife trade—cages with tightly packed populations of animals for sale—were quickly closed down, and the spread of the disease began to slow.

Scientists feel that China's reluctance to come forward to admit the outbreak was largely responsible for the crisis. Had the government been willing to address the disease outbreak among animals earlier, perhaps the outbreak, the fear, and the travel bans would never have had to occur.

Both SARS and avian and H1N1 flu (discussed later in the chapter) originate in wildlife reservoirs where parasites sometimes are pathogenic for certain species but do not necessarily damage the hosts. Ideally, the destruction of the animal population that is hosting the pathogen is the most effective way to halt the spread of the illness. Scientists stress the importance of identifying the reservoirs of these emerging pathogens so that they can understand how the diseases emerge and predict and prevent future outbreaks.

HIV/AIDS

Young people today, born after the initial AIDS epidemic of the 1980s, seem to accept that among certain populations in our modern society the HIV virus is simply one of the very real dangers of unprotected sex. Originally, of course, AIDS was a death sentence—killing relatively quickly and killing the young. Though there continues to be no cure and no vaccine has been fully successful, today pharmaceutical cocktails are keeping victims alive and comfortable enough that many have an acceptable quality of life, despite the illness.

However, if scientists of the 1970s and 1980s knew what they know now, AIDS could have been slowed or possibly eliminated in the human population. AIDS was first recognized in the United States in the spring of 1981, and there were at least 181 cases by the end of the year. It took another two years for scientists to trace the disease to the HIV virus, and they compared the DNA of HIV with the DNA of various related viruses. Researchers have identified SIV (simian immunodeficiency virus) in monkeys, and they think that the virus made the jump to humans when two different simian immunodeficiency viruses combined in the monkey population. Chimps sometimes kill and eat monkeys, and scientists speculate that the disease thus entered the chimp population, from which it finally made its way to humans through those who hunted and ate infected chimpanzees. The chimpanzee SIV became the deadly human immunodeficiency virus when it infected humans. If there had been a better understanding of animal-human disease transmission, along with a better disease surveillance system in the 1980s, perhaps AIDS would never have gained the strong foothold that it still has in the human population today.

WORRIES ABOUT AVIAN FLU

Concerns remain over the possible worldwide spread of the avian influenza (H5N1) that was very much in the news in 2005 and 2006 when the virus was spreading quickly through bird populations. Scientists feared that it would mutate into something that

An Indian veterinary staff member holds a pair of birds during a culling operation in a birdflu–infected area. *(Bapi Roy Choudhury for* Indigenous Herald *magazine)*

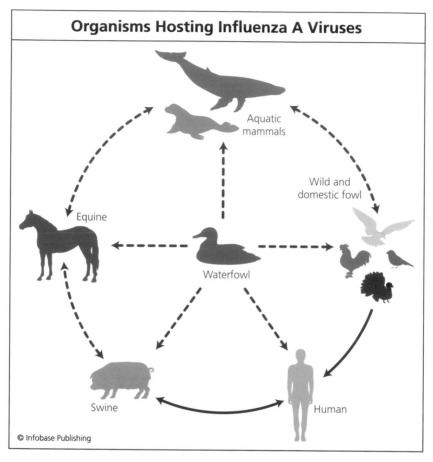

Organisms Hosting Influenza A Viruses

Aquatic mammals

Wild and domestic fowl

Equine

Waterfowl

Swine

Human

© Infobase Publishing

Some strains of influenza A virus can cause severe disease both in domestic poultry and occasionally in humans. The fear of viruses being transmitted from wild aquatic birds to domestic poultry raises the possibility of it becoming a human flu epidemic.

could transfer easily among people. Though a pandemic did not occur at that time, scientists still keep careful watch on this virus because the elements that led them to worry in 2006 are still in place. The spread of disease in the right environment could revive these fears. While the avian flu still presents a very real threat, the outbreak of the H1N1 (swine) strain of influenza A in the spring of 2009 has presented new worries. This outbreak and the possible scenarios it could follow will be outlined later in this chapter.

The H5N1 influenza (avian) virus is a type A deadly strain. (The name is an abbreviation for a scientific description of its structure.) The disease has been wiping out chickens and ducks, primarily in Asia, for a good number of years. The first transmission to a human occurred in Hong Kong in 1997, and to date the disease appears to kill about half of the people who become infected. In comparison, the seasonal flu that strikes each winter kills only a small fraction of those who become ill, still about 36,000 Americans.

Thus far, the virus is being spread via migratory birds that do not seem to fall ill right away, so they can fly for a few hundred miles after infection. Then, for a week or so they seem to shed the virus into the lakes or marshes where they land. Other wild birds become infected, fly away, and further spread the disease. Domestic fowl, who encounter the wild birds or the water where they have been, become ill. In the domestic birds, the virus seems to be more virulent and kills quickly; it is thought that the virus becomes more virulent in populations that live in close quarters.

Until spring 2005, the virus was contained to Southeast Asia. By July, infected birds had carried it to Siberia, where the north-south flyway meets the east-west flyway. As a result, it has now spread as far as Turkey and Iraq. Since 2003, the World Health Organization (WHO) has reported that patients with the H5N1 virus have been identified in Vietnam, Cambodia, China, Indonesia, Thailand, and, most recently, Turkey and Iraq, with the number of people infected and dying increasing each year. Thus far, there have been about 140 human cases in Asia, according to WHO, and at least 15 cases in Turkey. The official numbers of cases and deaths include only those confirmed by the WHO, making it possible that others have sickened and died without seeking medical treatment.

In January 2006, it was announced that a 15-year-old Iraqi girl died of bird flu after touching a dead bird infected with the disease. Because of Iraq's close proximity to neighboring Turkey, no one was surprised that the disease had spread there. However, flu experts note that this is their nightmare—that H5N1 enters the

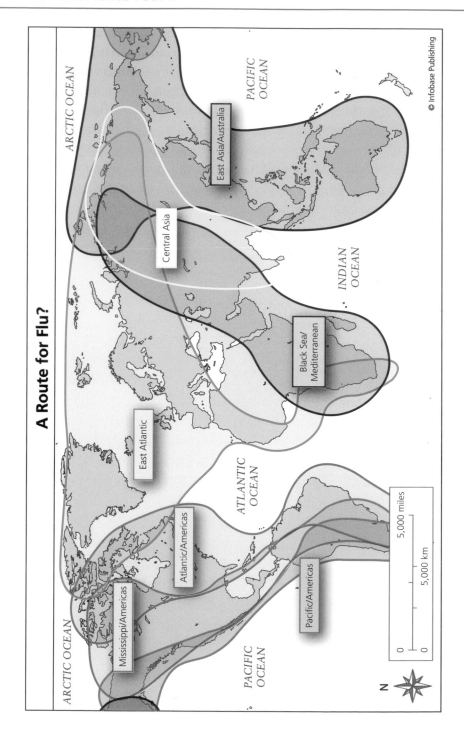

A Route for Flu?

ARCTIC OCEAN

ARCTIC OCEAN

PACIFIC OCEAN

PACIFIC OCEAN

ATLANTIC OCEAN

INDIAN OCEAN

East Asia/Australia

Central Asia

Black Sea/ Mediterranean

East Atlantic

Atlantic/Americas

Mississippi/Americas

Pacific/Americas

5,000 miles

5,000 km

0

0

N

© Infobase Publishing

human population before anything can be done. If the outbreaks are reported when it is still in the bird population, there are ways to try to stem the spread of the disease. Once it has infected an area long enough to spread to people, it is a great deal harder to control. For example, Turkey allowed the disease to travel throughout the country, so then it had to work to contain 55 outbreaks in 15 provinces.

To date, it appears that all those who have developed the deadly disease were in direct contact with infected ducks, chickens, or possibly pigeons. There is not yet any definite evidence that the virus is being transmitted person to person. But viruses mutate rapidly, and the fear is that the virus will evolve into a form that can be spread among people.

HOW THE VIRUS MIGHT MUTATE

Scientists know that there are two possible ways for the virus to become one that can be transmitted among people. The first occurs when avian and human flu strains combine genes. This genetic mixing can occur when a person is infected by both avian flu and a human flu strain at the same time.

The second possibility is that the virus itself may undergo enough *mutation* to make the leap. A recent analysis of the 1918 flu strain (See the sidebar "What Scientists Learned from the 1918 Flu Virus" on page 62) indicates that it may not take much for the flu to mutate on its own.

The scenario painted by the experts is only speculation—albeit educated speculation—that most world leaders are taking seriously. And, as always, there is concern that it could be a false alarm such as that which occurred with swine flu in 1976. Epidemiologists were very worried about the flu strain that was anticipated

_____

(opposite page) Scientists have traced what are known as flyways (the routes various species of birds use in migration) as the likely paths for the spread of avian flu.

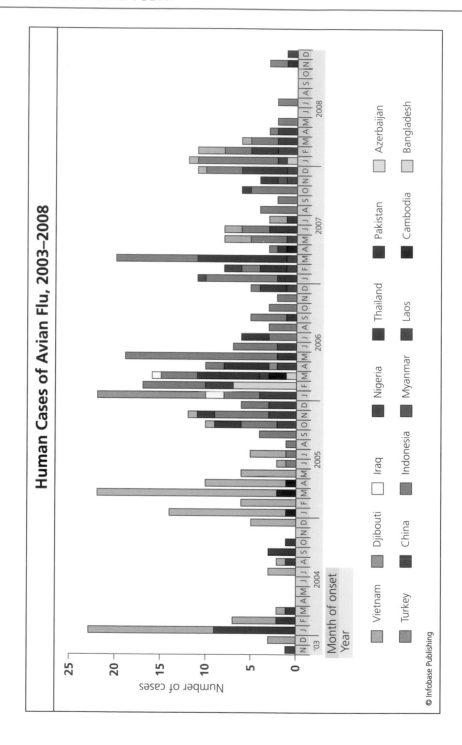

Human Cases of Avian Flu, 2003–2008

to spread that winter, and the U.S. government tried to prepare quickly for a nationwide vaccine program. Eventually one-third of all adults were vaccinated, but the swine flu epidemic never developed. By the conclusion of that flu season, more people suffered serious side effects from the vaccine—including the debilitating Guillain-Barre syndrome—than died from that winter's flu. This makes governments gun-shy. The difficulty is in assessing the risk and deciding if and when the avian flu is an active threat to the human population.

The Unexpected Outbreak of a Different Flu Strain

In June 2009, the World Health Organization (WHO) declared the spread of the influenza A H1N1 virus (or swine flu) to be a pandemic. This is the first time since 1968 that the WHO has upgraded the spread of a flu virus to a full level six alert. While the word pandemic creates fear for the public, to the WHO, this is a way of acknowledging the spread of the illness across a broad region and to let governments know that it is appropriate to step up the implementation of any plans to mitigate the spread of the disease. These plans could range from ramping up vaccine production to reviewing plans for community preparedness.

The strain of influenza that is currently of concern is influenza A virus subtype H1N1, which is described by scientists as a slightly different mix from what has occurred before. This strain is a combination of one type of virus that is endemic in humans, one that is endemic in birds, and two that are endemic in pigs (swine). With the spread of the disease to 76 countries in just a few months, the WHO stressed that the June 11, 2009, pandemic declaration was a result of the global spread of the virus, not its severity. Thus far, by mid-June 2009, only 180 deaths could be confirmed as attributable to this flu. However, scientists are concerned that the disease may become more virulent by next

(opposite page) This chart reflects the number of human cases of avian flu and the countries where they have been.

autumn. Even if it is no more devastating than seasonal flu, it could still be quite bad. On average, annual influenza is estimated to affect 5 to 15 percent of the global population, resulting in severe illness in 3 to 5 million patients and causing 250,000 to 500,000 deaths worldwide.

While the H1N1 virus is thought to have started infecting people as early as January 2009, it was not picked up as anything of note until late March when Mexico City first reported a surge in influenza-like illnesses that had begun occurring as early as February. The virus was not identified as a new strain of influenza until April 24, 2009.

As the disease spread throughout Mexico and into the United States, scientists began puzzling over a curious circumstance. In Mexico, the death statistics for this flu were far higher than those in the United States where most people have experienced it as a mild illness. Those Americans who have died thus far have usually had another underlying illness. Some scientists theorize that Mexico may actually have had a much larger epidemic of this flu than is being reported and that would explain the far higher numbers of those for whom the disease has been more serious or fatal.

Experts feel that the six years spent worrying about avian flu (first appeared in 2003) has helped prepared the United States and the rest of the world for the current outbreak. While experts feel the response has been better than it would have been 10 years ago, they note that there is still room for improvement. A nonprofit organization called Trust for America's Health has noted good progress on issues such as the stockpiling of antiviral medications (Tamiflu), the opening of new factories where vaccines could be made, and the fact that communities have made pandemic plans and conducted emergency drills. However, experts note that some of the other challenges of pandemic management remain problematic:

- The spring 2009 outbreak severely overwhelmed labs in the regions affected. Because there were so few places that could process the tests for the virus, many cases are suspected but not necessarily confirmed.

- Hospitals are working to refine their plans. People without health insurance often appear in emergency rooms with minor health concerns, and people who are mildly ill as well as the "worried well" divert hospital resources away from those who need it.
- The vaccination production capacity is still inadequate. Though creation of a vaccine is underway, having a brand-new vaccine ready in large quantities by autumn 2009 has been challenging.
- When schools closed in the spring of 2009 because of flu outbreaks (per community emergency plans), parents were caught between job and home demands, and there was no good solution.
- According to Trust for America's Health, 48 percent of working Americans do not get paid sick days. This means that many will try to come to work even if they are beginning to get sick, and this will increase the spread of the illness.

As fall 2009 approached, pharmaceutical companies worked to increase production of the vaccine they think will be effective against the virus, and countries are thinking through their emergency plans, while the world waits and wonders how serious the virus will be in the autumn.

Fighting the Flu

The challenge of an illness such as influenza is that it is not restricted to any particular population—anyone can catch the flu. Medical researchers know that a virulent virus can have a devastating effect on the world's population. While statistical reports from the Spanish flu that spread in 1918 are not considered fully accurate, they found that 50 percent of people came down with some form of the illness, and from 20 to 100 million people were killed worldwide. This is more than double the number killed in World War II. For this reason, world medical leaders take the threat of influenza very seriously.

The ideal for controlling an influenza virus such as H5N1 would have been to step in to control the disease before it became so well established in the bird or animal population. For example, this would involve eliminating the virus in domestic poultry overseas (vaccinate the birds or eliminate them before the virus spread to migratory birds) before it could mutate into a strain that passes among people. Early detection and rapid response to bird flu on a global scale could drastically cut the costs of dealing with a full-blown human pandemic.

Because epidemiologists fear that the avian virus is now widespread enough to eventually make the transition to becoming an illness that can spread from human to human, scientists are working on the next stage of planning for a possible outbreak. The two avenues to fight a virus are vaccinations (the better option) and antiviral medications. Neither method is simple or foolproof.

Despite the number of years that scientists have understood how to make vaccines, vaccines are not easy to create. Each year, the makers of seasonal flu vaccines have to take an educated guess as to exactly which flu they think will be circling the globe the following winter. Existing vaccines are powerless against new strains, so scientists must try to anticipate any mutations of the virus so that any vaccine created will fight the right bug. In summer 2009, the pharmaceutical companies were rushing a vaccine into production that is supposed to be effective against the H1N1 virus that is anticipated to continue to spread throughout the world.

The vaccine-manufacturing process is also challenging. Up until recently, the process involved injecting a flu virus into a fertilized chicken egg. This involved mixing a wild virus with a lab virus to assure that the pathogen would replicate within the egg so that it could be used as a vaccine. The egg method takes nine months from when a virus is identified as a risk to the distribution of a vaccine, and this certainly does not provide enough time for a world supply to be created.

Now scientists have a more rapid process for creating vaccines that employs reverse genetics. With this method, the live flu virus is stitched into loops of DNA called plasmids. The plasmids assemble into whole flu viruses in the lab. By using reverse genetics, scientists can create exactly the types of virus they need without having to mix viruses, such as adding the lab virus. They should be able to create the basic vaccine in a much briefer length of time (possibly as quickly as four weeks).

If there is not enough time to create a vaccine—or if the one created is ineffective—then doctors would turn to the antiviral medications. Two drugs, oseltamivir (commonly known as Tamiflu) and zanamivir (Relenza), may reduce the severity and duration of the illness. Both are from a class of medications that can inhibit the activity of the flu virus protein, but to be effective they must be started within 48 hours of symptom onset. Therefore, the people who start out with vague symptoms and ignore them may lose the window of opportunity to use an antiviral medication. In addition, antiviral medications become ineffective if overused, so there is concern among medical professionals that they may no longer be effective when they are ultimately needed.

While the possibility of any type of pandemic certainly catches the attention of both the press and the public when scientists sound the alert, it is interesting to look back and see that other illnesses have also caused this same type of alarm. On July 13, 1883, the *New York Times* ran a notice about the spread of cholera in Egypt, and that article was accompanied by stories from both Paris and London. In Paris, the Hygiene Commission agreed that Louis Pasteur should journey to Egypt to offer help; in London, Mr. Gladstone announced to the House of Commons that the government would send a British surgeon-general who had experience in treating cholera. History seems to repeat itself.

HOPE FOR THE FUTURE—CONSERVATION MEDICINE

Scientists have learned that damaged ecosystems—characterized by toxins, degradation of habitat, removal of species, and climate change—have created conditions for pathogens to move in unexpected ways, including from animals to people. The Wildlife Trust, a global organization focused on conservation science and the link between ecology and health, created the Consortium for Conserva-

WHAT SCIENTISTS LEARNED FROM THE 1918 FLU VIRUS

The H5N1 influenza virus is closer in composition to the Spanish flu of 1918 than it is to other viruses that have circulated, so for that reason the 1918 virus has been of interest.

The Spanish flu seems to have begun as a normal human flu strain that circulated in the United States in the spring of 1918. By summer, the virus had reached the battlefields of World War I. There, among people packed into the trenches, trucks, trains, and hospitals of the western front, it turned lethal, just as the avian flu does among birds on crowded chicken farms today.

Because scientists in the early 20th century had not learned to isolate viruses for study, there had been no data and no simple way to learn any lessons from that pandemic. Dr. Jeffery Taubenberger, chief of the molecular pathology department at the Armed Forces Institute of Pathology in Washington, felt that the Spanish flu had stories to tell, so he developed a plan to recreate the virus in order to study it. In a 10-year project completed in the fall of 2005, Taubenberger and his team isolated and began to study the Spanish flu.

Working under extremely secure conditions so as not to endanger the staff or permit the escape of the virus, Taubenberger and his research group located lung tissue from two soldiers who died of the 1918 flu from an autopsy tissue warehouse originally established by President Lincoln.

tion Medicine (CCM). This collaborative institution links Wildlife Trust with Johns Hopkins Bloomberg School of Public Health, Tufts University School of Veterinary Medicine Center for Con-

The importance of careful handling of the virus was brought home very recently when some testing on the SARS virus in Beijing resulted in several laboratory workers becoming infected with the virus from working with it.

A third sample was sent to him by the retired pathologist Johan Hultin of San Francisco, who spent his own money to go to Alaska where he gained permission to excavate a body from a mass grave in a community where 72 adults died from the flu in 1918. Because the grave was dug in permafrost, the bodies had remained frozen all this time. He snipped frozen lung tissue from a woman and sent it off to Taubenberger for his study. Taubenberger himself was busy procuring samples of birds that had died in 1918 and 1919 from the Smithsonian so that he could identify for certain that this was a bird virus.

Though it took a full decade to piece together the virus, researchers have not only analyzed the virus, but they have also conducted gene-swapping experiments to determine what weakens the virus and what makes it stronger.

In comparing the 1918 flu with today's human viruses, Taubenberger found that the virus had alterations in just 25 to 30 of its amino acids. Those few changes—much fewer than expected—turned a bird virus into a killer virus that could spread from person to person. The current avian influenza virus has already made 5 of the 10 changes found in the 1918 virus.

These hands holding a globe depict how very interrelated the entire world has become. *(Paradise Ridge Chamber of Commerce, Paradise, California)*

servation Medicine, University of Pittsburgh Graduate School of Public Health, University of Wisconsin–Madison Nelson Institute for Environmental Studies, and U.S. Geological Survey (USGS) National Wildlife Health Center (NWHC). These scientists and medical personnel combine their efforts to interrupt transmission routes and prevent future disease outbreaks.

Close to home, scientists point to what has happened with Lyme disease, which was first identified in Old Lyme, Connecticut, and therefore bears that town's name. While the pathogen that causes Lyme disease has been around for a long time, it did not create a problem for humans until recently as forests have been chopped down to make room for suburban homes.

Lyme disease is carried to humans primarily via ticks that often travel on white-footed mice. Mice seem to be perfectly happy living in the suburbs, and as they become infected by the ticks that suck their blood the reduction of forests has meant that there are fewer wild animals to eat the mice to keep down the Lyme disease–carrying mouse population. Ticks in the suburbs are more likely to live in infected mice, and when they take their next meal on a human or the family dog, the result has been the explosive spread of Lyme disease.

A similar interference with nature has happened as an increasing amount of the Peruvian rain forest has been destroyed. There has been an explosion of malaria-bearing mosquitoes that thrive in the sunlit ponds created by logging operations.

THE DEADLY DOZEN FROM THE WILDLIFE CONSERVATION SOCIETY

Founded in 1845 and previously named the New York Zoological Society, the Wildlife Conservation Society (WCS) was one of the first U.S. conservation groups. It has grown and expanded and now manages the world's largest system of urban wildlife parks. It consists of a worldwide network of scientists whose goals are not only to safeguard animal habitats in danger of being lost to civilization, but also to be pioneers in the field of wildlife-human disease threats.

One of the first diseases that scientists realized made the animal-human species jump was AIDS, and since that time the WCS has been sounding the alarm that climate change is causing the spread of pathogens into areas in which they have never existed before.

The diseases about which the WCS is most concerned are what they call the deadly dozen:

- avian influenza (previously discussed in this chapter)
- babesia. This disease is carried by ticks and infects domestic animals and wildlife. Animals who sustain a particularly large number of tick bites are shown to be susceptible to other illnesses. In East Africa, there was a huge die-off of lions from distemper, but the initial weakness seemed to come from the infection from tick bites. More and more humans are developing babesia.
- cholera. This disease is waterborne and results in severe diarrhea. It mainly affects people in developing countries, but because cholera is more likely to occur in warm climates, the trend toward warmer climates throughout the world makes scientists worry that this will spread.
- Ebola. This is a hemorrhagic virus that kills humans, gorillas, and chimpanzees, and there is no known cure. Ebola outbreaks seem to come during variations in seasonal rainfall. Thus, as the climate changes, scientists

expect more outbreaks. They also anticipate that the areas where there are outbreaks may expand.

- parasites. Warmer temperatures create more environments in which parasites can survive. Many types of parasites travel easily between humans and wildlife, so monitoring parasites in animals becomes even more vital to protect human health.

- plague. This infectious disease has been around for centuries. It is generally spread by rodents; fleas attach to the rodents and then spread the disease to humans. It is expected that rodents will begin to live in wider areas, thus extending the range of rodent-borne diseases.

- Lyme disease. This has become increasingly prevalent as humans have encroached on wildlife habitats. The disease is transmitted to humans through tick bites. As tick season extends and their range expands, there is concern

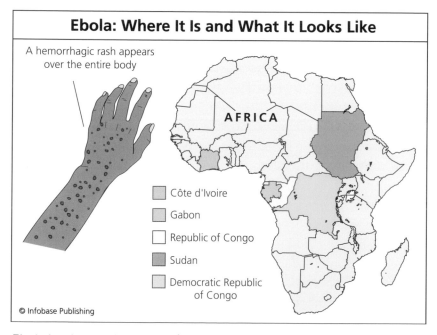

Ebola: Where It Is and What It Looks Like

A hemorrhagic rash appears over the entire body

AFRICA

- Côte d'Ivoire
- Gabon
- Republic of Congo
- Sudan
- Democratic Republic of Congo

© Infobase Publishing

Ebola is a hemorrhagic fever virus that has crossed over from animals to humans. In most forms it is very deadly.

that this neurological illness will be communicated to more people.

- red tides. In warmer waters, red algae creates toxins that are deadly to both humans and wildlife. These occurrences—commonly called red tides—cause mass fish kills, marine mammal strandings, and death to penguins and seabirds, as well as human illness.
- Rift Valley fever. This illness affects public health and can be spread via food contamination. It is particularly common in Africa and the Middle East. It travels via mosquito.
- sleeping sickness. Both animals and people are affected by this illness that is endemic in parts of sub-Saharan Africa. It is carried by the tsetse fly.
- tuberculosis. The big concern today is about what is known as bovine tuberculosis. As cattle are transported to various parts of the world, the disease has spread, and it can be communicated to people via unpasteurized milk.
- yellow fever. This illness that created so much trouble in the late 19th century in Central America (and actually North America) is still very much alive. One type of yellow fever—what is known as jungle yellow fever—spreads from primates to humans easily.

These are just a few of the diseases that have made the animal-human species jump. Scientists will be monitoring wildlife to see what other illnesses gain the ability to spread from animals to humans.

CONCLUSION

Scientists today are highly aware that they need to vigilantly watch for disease outbreaks in both the animal and human populations of the world. Disease surveillance is costly, and destruction of domestic animals is very unpopular as animals or birds that appear healthy often must be sacrificed simply because they have been exposed to those with a deadly illness. As a result, there is a major effort for organizations such as the WCS and the WHO to do what they can in the way of global surveillance and outreach. Curtailing the spread of new and emerging diseases is far preferable to fighting them once they have invaded a population where they are very likely to be transported to all parts of the world in today's global economy.

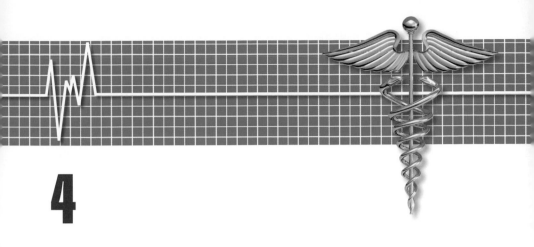

4

Medical Frontiers of the Future

Scientists throughout the world are working on possible cures for human disease, ranging from the common cold to *autoimmune* diseases (where one's body launches what can be devastating attacks on itself). There is much progress under way, but for the purposes of this book this chapter will outline just a few of the methods that have been making a great deal of news recently—the use of stem cells, genetically created medicines and diagnostic techniques, and nanotechnology.

For many researchers, stem cells offer the promise of being the basis for health therapies that today are only dreamed of. The actor Christopher Reeve (1952–2004) was paralyzed by a fall from a horse in 1995, and he always felt that stem cells would be the answer to letting him walk again. While there is a lot of hope for stem cell research, the process holds no guarantee.

Physicians today predict that one day genetics will permit tailoring both diagnoses and treatments for individuals based on their personal genetic profiles. When this development comes to pass, it will make the generalized form of medicine used today seem as antiquated as bloodletting.

Nanotechnology, the world of microscopic creations, holds great promise for medicine, but it is not without its risks. This chapter will explain how nanomedicine might be used, but it also outlines paths that must be explored first. There may be hazards to these micro-creations, and scientists will need to understand what could be at stake before they proceed with any implementation.

THE HOPE FOR STEM CELLS

In the early 1960s, the Canadian researchers J. Till and E. A. McCulloch began working on a process that involved injecting bone marrow cells into mice that had undergone radiation. They observed that nodules appeared in the mouse spleens in direct correlation to the number of bone marrow cells injected. Till and McCulloch began to suspect that each nodule arose from a single marrow cell—perhaps a stem cell. They were joined in their work

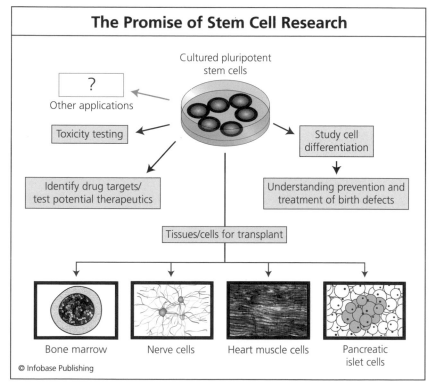

The Promise of Stem Cell Research

Cultured pluripotent stem cells

?
Other applications

Toxicity testing

Study cell differentiation

Identify drug targets/ test potential therapeutics

Understanding prevention and treatment of birth defects

Tissues/cells for transplant

Bone marrow   Nerve cells   Heart muscle cells   Pancreatic islet cells

© Infobase Publishing

Scientists continue to hold great hope for the promise of how stem cells can be used to rebuild or treat parts of the body.

by a graduate student Andy Becker and a molecular biologist Lou Siminovitch, and together the group determined that these cells were capable of regenerating. This was a very exciting concept. Further gains were made by James Thomson at the University of Wisconsin who was the first scientist to create a stable line of human embryonic stem cells. In 1998, he removed cells from inside a week-old embryo (*blastocyst*) and *cultured* these cells, 50 to 100 of them, in a laboratory dish.

The discovery of stem cells has brought great excitement to medical researchers because stem cells are unspecialized cells that have two primary and very important properties:

1. They can differentiate and become other cells in the body.
2. They have the ability to regenerate on their own.

In addition to having the ability to develop into many types of specialized cells within the body, stem cells seem to be able to divide without limit. They seem to serve as a sort of repair system for the body. The new cell created may remain a stem cell or it may go off to serve a function where it is most needed—perhaps becoming a muscle, blood, or brain cell. Scientists are working to determine how and whether they can control this differentiation process as that would offer amazing prospects for developing new types of cures.

The Process

After scientists remove an inner cell mass (ICM) from a blastocyst (usually one left over from an attempt at in vitro fertilization), the cells are placed on a plate with feeder cells. Scientists have found that cells from mouse skin work particularly well for this. A *totipotent* stem cell comes from a fertilized egg and can develop into any type of cell in the human body. A *pluripotent* stem cell is slightly more limited. The ability to divide and produce more cells is key. If all goes well in the process, then in a few days new cells grow and form colonies. These cells are determined to be embryonic stem cells only if they display certain

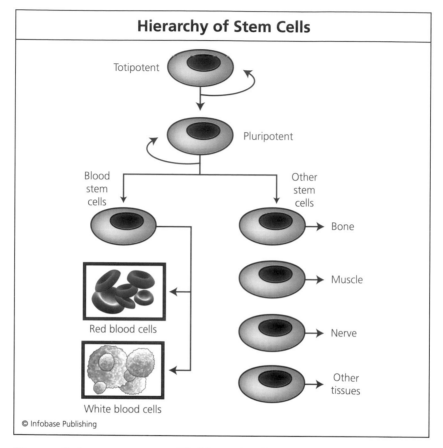

Hierarchy of Stem Cells

Totipotent

Pluripotent

Blood stem cells

Other stem cells

Bone

Muscle

Nerve

Other tissues

Red blood cells

White blood cells

© Infobase Publishing

Stem cells have a very specific hierarchy that depicts how they can be used. The top level, the totipotent cell, has the potential to become any type of cell. The other cells do not have quite the breadth of possibilities as do the totipotent cells.

molecular markers agreed upon by scientists and go through several generations of cell divisions.

Stem cells in an adult are present in lesser numbers than they are in embryos, though bone marrow in all people is filled with stem cells. The role of adult stem cells (also called somatic stem cells) is generally believed to be in repair of damaged and injured tissue. As early as the 1960s, Till and McCulloch developed a cell-cloning technique that permitted bone marrow transplants for diseases like leukemia and other blood disorders.

Many scientists feel that embryonic stem cells are a better source for further medical research and application, as these cells have not yet become specialized in any way. (Parents today are now encouraged to have stem cells from their newborn's umbilical cord frozen for possible use later for to-be-discovered medical cures.) There is particular hope that stem cells will be helpful in managing degenerative illnesses such as Parkinson's disease. Neurological diseases like this are so complex that drugs, or even gene therapy, often prove to be inadequate. Stem cells offer hope.

Only 22 stem cell lines have been available for federally funded research in the United States, and during George W. Bush's administration it was mandated that the National Institutes of Health (NIH) should not support work on lines created after August 2001. (Once created a stem cell line can be kept going perpetually through freezing and storage.) One of President Barack Obama's early actions in 2009 has been to loosen these restrictions, though scientists had already begun to create new ways to make stem cell lines.

Because of the Bush administration rulings, scientists had been looking for alternative ways to gain stem cells. In 2005, scientists at Harvard succeeded in turning ordinary skin cells into what appear to be embryonic stem cells—without having to use human eggs or make new human embryos. This was a big step forward in gaining access to stem cells without having to work with embryos, which stir up such controversy. A possible benefit of the technique is that the cells come directly from the patient, so the DNA in the new stem cells is an exact match. This could enhance the chances of a person's body more readily accepting whatever it is the stem cells need to be used to replace.

Embryonic stem cells must be directed to specialize before they are used. In experiments with mice, scientists saw that direct introduction of embryonic stem cells can cause cancer. If stem cells were to be used to produce insulin for diabetics, for example, it would be vital that the stem cells had been fully converted and that there were absolutely no stem cells remaining before they were introduced into a patient. Scientists are just learning what variables and nutrients are necessary to accurately guide stem cells so that they become stable nerve or muscle cells.

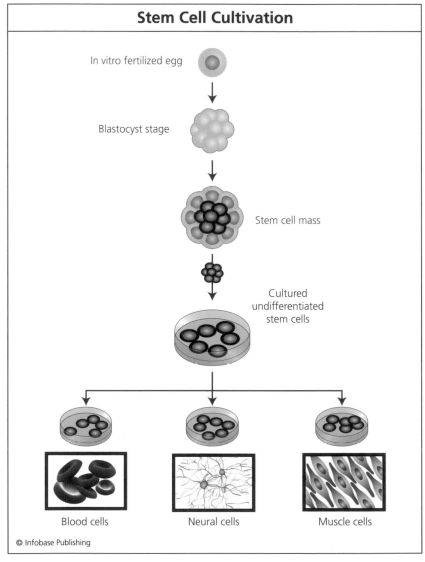

Stem Cell Cultivation

In vitro fertilized egg

Blastocyst stage

Stem cell mass

Cultured undifferentiated stem cells

Blood cells

Neural cells

Muscle cells

© Infobase Publishing

Scientists are experimenting with how to cultivate stem cells in order to learn more about biology and how the stem cells may eventually be used to improve human health.

In addition to the controversy over stem cells (discussed in the following section), there are still many obstacles to using stem cells. Scientists need to develop better and more reliable ways to obtain embryonic stem cells, and they need to learn to control the way they differentiate. Most countries are struggling similarly to the United States with the ethical issues surrounding stem cells, reproductive cloning, and therapeutic cloning. (See the sidebar "How a Cloned Sheep Could Help Medicine" on page 78.) In the United States, there has been some activity in various states to permit the research before the federal government did. California was one of the first states to break with the Bush sentiment on stem cell research.

Though no one yet understands exactly how the stem cells can be put to work, the following are some of the ways it is anticipated that stem cells might be used:

- to further studies of genetics
- to shed light on inherited diseases, particularly degenerative ones
- to correct genetics in children. For example, if a child has poor immune response to certain stimuli, this may be corrected by working on a specific gene.
- to create cell-based therapies
- to study biological processes
- drug discovery and development
- to repair spinal cord injury

Amazingly, stem cells have already been featured in scams. Patrons of Russian beauty parlors have been told that stem cell injections can be given to them, and promises are made of a long list of ills the process can cure.

THE CONTROVERSY OVER STEM CELLS

The early embryo from which stem cells are derived is a ball of cells smaller than a grain of sand. While it could become a person,

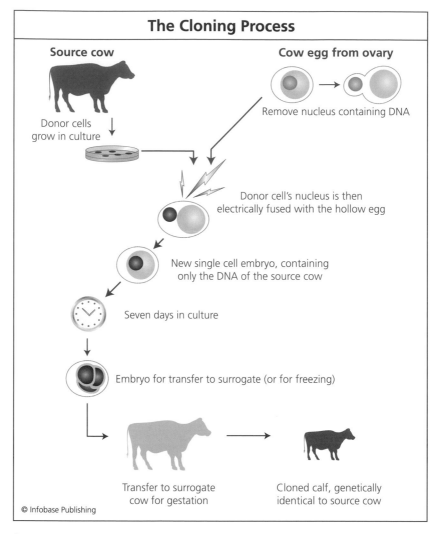

The Cloning Process

Source cow

Donor cells
grow in culture

Cow egg from ovary

Remove nucleus containing DNA

Donor cell's nucleus is then
electrically fused with the hollow egg

New single cell embryo, containing
only the DNA of the source cow

Seven days in culture

Embryo for transfer to surrogate (or for freezing)

Transfer to surrogate
cow for gestation

Cloned calf, genetically
identical to source cow

© Infobase Publishing

Scientists are still experimenting with how and when cloning might be
safe and beneficial.

it lacks any sort of human characteristics at this stage. However,
many people are deeply offended at the thought of scientists working with a human embryo. There are legitimate religious and ethical concerns. Some feel embryos should be given the protection of
human rights. Certainly, an informed debate is in order. Just as
there are difficulties in deciding about when it is permissible to

remove organs from accident victims, there are similarly no easy answers to this type of research.

Politicians and the public are most concerned about the direct use of embryonic stem cells being placed in other human beings, but scientists note that in the long run stem cells may be most valuable in the laboratory. If researchers can use stem cells to work out the complex steps involved in a medication's effect on the body, for example, it may prove that the major benefits are simply for research. Or if scientists were to determine that stem cells carried the genetic markers for cystic fibrosis, then this would permit them to explore possible cures using only the cells.

This debate is far from over. Scientists still need to determine exactly how stem cells could be of the most value, and in the meantime there will continue to be discussions about guidelines governing their use.

GENETIC TESTING AND GENE-BASED MEDICINE

By applying genetic science to health issues, scientists foresee a day when it will be easier to diagnose, manage, and counsel individuals with hereditary illnesses. Genetic medicine is the catchall phrase used when discussing gene therapy, personalized medicine, and the rapidly emerging new medical specialty, predictive medicine.

Gene Therapy

Genes are carried on *chromosomes* and are the basic physical and functional units of heredity. When genes are harmed so that the encoded proteins are unable to carry out their normal functions, genetic disorders can result. Gene therapy is a technique for correcting defective genes responsible for disease development. Researchers may use one of several approaches for correcting faulty genes. Target cells such as the patient's liver or lung cells could be infected with a viral agent that could unload its genetic material containing the therapeutic human gene into the target cell. The generation of a functional protein product from the

(continues on page 80)

HOW A CLONED SHEEP COULD HELP MEDICINE

In 1997, two scientists did the unthinkable—they cloned one mammal using the cells of another adult animal. Cloning had previously been performed using embryonic cells, but this was the first time it had been performed using cells from an adult animal. The scientists involved were Ian Wilmut and Keith Campbell, working with a staff at the Roslin Institute in Edinburgh, Scotland. The animal they cloned was a sheep named Dolly, who became famous throughout the world. Dolly lived to be six years old.

The cell from which Dolly was cloned came from the mammary gland of another sheep; the concept that a full

Dolly the sheep *(Wikimedia)*

additional being could be created from a cell taken from a specific part of the body was very instructional for scientists.

In the process, scientists learned that a somatic cell (adult cell) was capable of reverting back to an undifferentiated totipotent state and could create cells that went on to create a full being. Thus far, this method is far from exact—the process still produces its misses, and many of the embryos that have been cloned have developed abnormally. Out of 277 attempts, Dolly was the only lamb to live to adulthood. Though other animals have been cloned since Dolly, Ian Wilmut, who most recently has been chair in reproductive science at the University of Edinburgh, made a statement in 2007 noting that the process was still highly experimental and should not be undertaken lightly.

The cloning of Dolly in 1997 raised the possibility that one day a human might be cloned. While the full cloning of an individual raises many questions that would need a great deal of ethical discussion and oversight, the discovery opened the possibility of therapeutic cloning, which might be helpful in solving medical problems. Therapeutic cloning is a process used to create embryos as a source of embryonic stem cells for therapeutic purposes. Because embryonic stem cells can grow into any type of cell, they might be cultured into nerve cells, skin cells, or whatever kind of cell is needed. The obvious use of therapeutic cloning would be treating deadly diseases like diabetes and Parkinson's, where a specific type of cell has died. Replacing those cells would be expected to restore health to an individual.

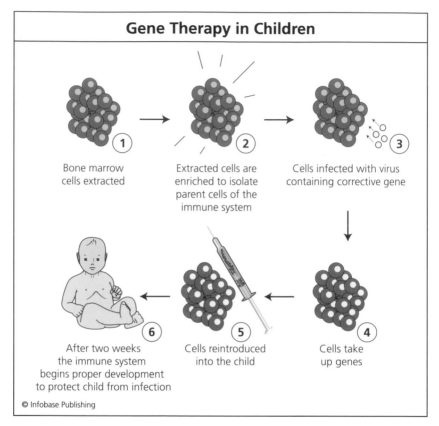

Gene Therapy in Children

1 Bone marrow cells extracted

2 Extracted cells are enriched to isolate parent cells of the immune system

3 Cells infected with virus containing corrective gene

4 Cells take up genes

5 Cells reintroduced into the child

6 After two weeks the immune system begins proper development to protect child from infection

© Infobase Publishing

It is hoped that one day gene therapy can be used to cure children who have chronic or debilitating illnesses.

(continued from page 77)

therapeutic gene restores the target cell to a normal state. While there is great hope for this type of treatment, there are still many safety issues to work out. In addition, scientists are trying to figure out ways that gene therapy can provide long-term benefits; right now most benefits are short term.

Personalized Medicine
Patients and doctors can testify that no medicine works the same way for different individuals. Because of this, scientists are trying

to develop ways that they can use a person's genetic profile to create a medicine that would be specific to each individual.

Pharmacogenomics describes the study of how individuals react to medications. A few tests are already available that can detect some of these genetic variations and predict how a person is likely to respond. Scientists are trying to identify and record as many genetic variations as possible. Once a variation is identified, scientists might be able to match it up with a response to a particular medication and then develop a personalized approach to creating medicine. Doctors will one day take into account such factors as a patient's weight, age, medical history, and even how any blood relatives have reacted

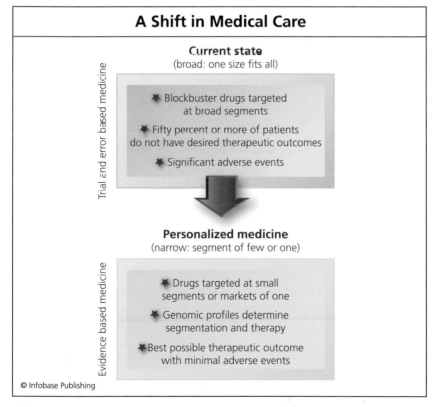

The current state of medicine consists of pharmaceutical cures that are targeted for use by all. As genetic profiles become a more important part of medicine, prescriptions will become very personalized.

to the same medication. While there would still be no guarantees, pharmacogenomics could take more of the guesswork out of the process, and the medicine provided to each individual would be more likely to be effective and have fewer side effects.

Predictive Medicine

The goal of predictive medicine is to evaluate a patient's genetic profile in order to predict future disease. Based on this information, health care professionals and patients can make lifestyle modifications to help avoid a particular health problem. For example, if high cholesterol runs in the family and a patient has genetic markers for it—or early signs of disease—eating habits can be adjusted. If this is not effective enough, then the physician would be aware of the family tendency and might medicate sooner than he or she would with another patient.

Zebrafish are scientists' preferred organisms for genetic study. They have been identified—and actually celebrated—for their ideal use for studying vertebrate development and gene function.

A zebrafish under study at Pennsylvania State University *(ZFIN and Oregon Zebrafish Laboratories, National Science Foundation)*

NANOTECHNOLOGY AND NANOMEDICINE

Nanotechnology describes the science of working in the world of the extremely tiny—the world of the single atom or the small molecule. To put this in perspective, a nanometer is a billionth of a meter, or a millionth of a millimeter. Scientists predict that within the next 50 years, machines will get smaller and smaller—to the point that thousands of these tiny machines would fit into the period at the end of a sentence.

The science itself involves condensed-matter physics, engineering, molecular biology, and chemistry. Nanoscience has already contributed greatly to many industries—there are products and processes in microprocessor manufacturing, heavy equipment manufacturing, and the aerospace industry. Nanoscience has provided the ability to make new catalysts, coatings, paints, and rubber and tire products as well. Nanoscience is also being used in skin preparations; sunscreens contain nanoparticles of titanium dioxide, which refracts light. If a sunscreen is manufactured with bigger particles, then the sunscreen appears white.

If atoms are nature's building blocks—and are the ingredients of everything from a human to a tree—imagine what might be possible if scientists learn to work at this scale for the good of humanity.

Nanotechnology Introduced

The distinguished physicist Richard Feynman first presented the concept of nanotechnology, but not the word, in a speech he gave to the American Physical Society on December 29, 1959, entitled "There's Plenty of Room at the Bottom." Feynman envisioned the day when people would have the ability to manipulate individual atoms and molecules so that one set of precise tools could be created to make a proportionally smaller set, on down to the needed scale.

Nanotechnology was not used until 1974, when Norio Taniguchi, a professor at Tokyo Science University, introduced it in a paper. The science got more fully underway when Gerd Binnig and Heinrich Rohrer created the scanning tunneling microscope

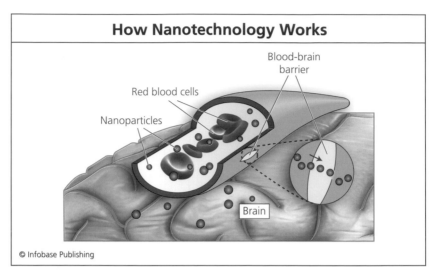

How Nanotechnology Works

Blood-brain barrier

Red blood cells

Nanoparticles

Brain

© Infobase Publishing

One day scientists hope that nanoparticles will be able to be sent into the body to bring about specific remedies, but many hurdles must be overcome before they are let loose. This graphic shows carbon nanoparticles that cross the blood-brain barrier and encourage clotting. If this can be appropriately directed, it is a good thing; if not, it could create other health problems.

in 1980, and individual atoms could actually be seen. During the next few years, Dr. Eric Drexler, now regarded as the father of nanotechnology, wrote *Engines of Creation: The Coming Era of Nanotechnology,* a book that more fully developed the concept.

One imperative of nanotechnology is that nanodevices must be able to self-assemble. Because these machines will be so small, it will literally take millions of them to do some tasks on a scale that would be helpful to humans. At Cornell they have created little four-inch-square (10.16 cm) blocks that can replicate themselves. The plastic cubes sliced diagonally into two halves that swivel, and the electromagnets on the cubes turn on and off, allowing them to pick up and release other cubes. A three-cube stack takes a minute to copy itself.

One concern about nanotechnology is that if it is created to self-build, the nanomachine might keep building and building and building unchecked. With this particular device, the Cornell scientists have developed safety features so that the machine cannot go unrestrained. To begin with, the machine relies on power from a base plate, and researchers must also feed the robots with new blocks. Another safeguard is that the number of blocks cannot exceed the number of plates.

THE GOALS OF NANOTECHNOLOGY

Those who foresee a world employing nanotechnology imagine a process that will involve anything from a nanomachine that can create a baseball one atom at a time to another type of nanobot that can speed through one's arteries, cleaning up cholesterol as it tools along. The goal must be to learn how to successfully manipulate material at the molecular and atomic levels, using both chemical and mechanical tools. There have been some successes, and some scientists predict that we could see definite progress within only 15 to 20 years.

Atoms can now be manipulated, separated, and put back together in different formations. (Atoms and molecules stick together because they have complementary shapes that lock together or charges that attract.) As millions of these atoms are pieced together by nanomachines, a specific product begins to take shape. To create nanotechnology-produced goods, scientists must learn to further manipulate individual atoms and create assemblers. Because the scale of this technology is so tiny, trillions of assemblers will be needed to do the work, so scientists will have to be able to make nanoscopic machines, called replicators, that will be programmed to build more assemblers.

Nanobots, perhaps what the machines will be called, will not simply be scaled-down versions of contemporary robots. The different physics at these scales means that human-made nanodevices will probably bear a much stronger resemblance to nature's nanodevices and will be made from proteins, DNA, and membranes—much like viruses.

Life figured out nanoscience long ago—each one of us has billions of molecular motors crawling around in every cell of our bodies right now. The key is figuring out how to direct human-made ones and keep them going.

The Hope for Nanotechnology

Under President Bill Clinton, the government doubled its investment in research and development of nanotechnology and termed it the new frontier. This work will affect many government agencies: the National Science Foundation, the Department of Defense, the Department of Energy, the National Institutes of Health, the National Aeronautics and Space Administration, and the National Institute of Standards and Technology. While this book focuses on ways nanotechnology will be used in medicine, scientists in other fields have great hope for this research, and they anticipate building materials stronger than ever. Also, as computerization becomes smaller and smaller, the day will come when computers can be built into such items as clothing.

Research will have a huge impact on the medical industry. Consider the following issues:

- Nanobots, controlled by nanocomputers or ultrasound, will be used to manipulate other molecules, destroying cholesterol molecules in arteries, destroying cancer cells, or constructing nerve tissue atom by atom to end paralysis. Patients might ingest the nanobots in fluids, making it a very simple medical process.
- Nanosurgeons may work at a level 1,000 times more precise than the sharpest scalpel, and there should be no scarring.
- New biomedical solutions for chronic disease will be developed.
- New drugs and targeted drugs for delivery will be created.
- Nanoscale medical diagnostic devices will be invented.
- Though ethics debates will abound, nanotechnology could even change physical appearance by rearrang-

ing the atoms of your nose or shifting someone's eye color.

- Nanotubes could be used for bone breaks. Scientists are working on a variety of ways to use carbon nanotubes, pencil-shaped molecules that are very strong but difficult to manipulate, because each tube is a tiny fraction of the width of the period at the end of this sentence. While many of the nanotubes will be used in manufacturing, two teams of researchers are working on medical applications. Because the human body can absorb carbon, scientists at Stanford University have created cancer-killing nanotubes designed to invade tumor cells. At the University of California, Riverside, they are experimenting with ways to use nanotubes to help heal broken bones.
- People in hazardous environments will have greater protection. Clothing will constantly monitor physiological vital signs, warn about exposure to chemicals, adjust for environment stresses, provide camouflage that matches changing background and lighting conditions, and even provide first-aid casualty response.

As noted in chapter 2, there is an unbreakable link between the environment and human health, and nanotechnology may have a tremendous impact in helping improve health by cleaning up the environment:

- Airborne nanobots could rebuild the thinning ozone layer.
- Contaminants could be automatically removed from water sources, and oil spills could be cleaned up instantly. Nanotubes with fingers 50,000 times thinner than a human hair would manipulate the atoms in an oil spill to render it harmless.
- Nanotechnology will pollute less, and we will reduce our dependence on nonrenewable resources.

Nanotechnology offers great promise for progress, but scientists are aware that there are risks involved. The risk-benefit analyses for nanotechnology will be vital to moving forward.

NANOMEDICINE—NOT WITHOUT RISKS

Dr. Eric Drexler, the father of nanotechnology, almost certainly regrets his early speculation that one of the slight risks of nanotechnology is that the self-replicating feature could lead to a situation in which nanobots run amok and take over the world—what is now called global ecophagy (literally, eating the environment). A variation on this fear is the green goo theory, in which nanobiotechnology creates a self-replicating nanomachine that consumes all organic particles, creating a slimelike nonliving organic mass.

The Risk of Green or Gray Goo

In a worst-case scenario, all of the matter in the universe could be turned into goo (goo meaning a large mass of replicating nanomachines lacking large-scale structure; gray goo refers to runaway nanobot self-replicators; green goo refers to an organic replicator as described above), killing the universe's residents. In the late author Michael Crichton's recent novel *Prey,* a company in Nevada accidentally/purposely releases self-assembling nanobots into the desert; they quickly replicate and evolve and threaten all the human protagonists.

According to Chris Phoenix, director of research of the Center for Responsible Technology, runaway replicators could be created only by a deliberate and difficult engineering process and not by an accident. Drexler adds that the scaremongering detracts from the promise of nanotechnology. But with the risks in mind, scientists are programming nanomachines to rely on an energy source or to stop reproducing after a certain number of generations. That way a nanomachine could not go unchecked. Bacteria have never taken over the world, so nanoparticles probably will not do so either.

Poison and Toxicity

Some of the most promising nanoscale materials have been found to be harmful in certain situations, so scientists will want to proceed cautiously. A high concentration of carbon nanotubes can fatally clog the lungs of rats, and fullerenes (spherical carbon cages just a few nanometers across) tend to accumulate in fish brains.

Nanoparticles in drinking water could be dangerous to humans or animals, and this new class of nanosubstances will need to undergo additional safety testing.

CONCLUSION

Stem cell research, genetic medicine, and nanomedicine all hold great promise for future gains in medicine. But like all other steps forward, it is vital that science approach these new methods carefully, with full awareness of the risks involved. If, after careful study, one or more of these methods proves beneficial, then the world will be the richer for the work involved in all the experiments and trials.

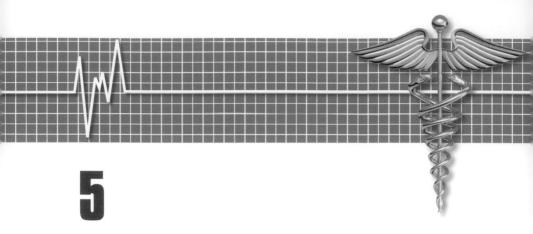

5

Modern Medicine and Medical Ethics

Ethical dilemmas for healers have existed since the beginning of time. To treat or not to treat was a difficult question for thousands of years, because healers—and later academically trained physicians—had no guarantee of what would happen to a patient; they rarely had any statistical research that would provide data that might be helpful in deciding what to do for a sick patient. In childbirth, midwives constantly faced the issue of losing the mother but saving the baby or vice versa. When physicians began to be on the sidelines of the battlefield, the decision as to who to treat first often dictated the fate of the injured men. Until military surgeon Dominique-Jean Larrey (1766–1842), priority treatment was given to officers—a triage method that would not be tolerated today.

While physicians of the past have long faced vexing problems, the medical profession today has an even more daunting range of dilemmas. As science continues to expand the medical possibilities, physicians are faced with enormous ethical questions for which there are no easy answers. As cloning and genetic manipulation and all types of transplants—the latest being face transplants—become possible, doctors will have to reckon with whether they should just because they can.

And as medicine makes it possible for the aged to become even older, with many attendant health problems, is science prepared to help out with quality of life issues? The person who is treated for cancer at the age of 92 who survives may be so debilitated by other health issues or by the cancer treatment itself that his or her quality of life is severely affected. In addition, the cost of the cancer treatment and future hospitalizations are all covered by Medicare, the government funding of health care for those over 65 years of age. Hospital administrators need to weigh in on whether financial considerations are ever a part of the medical decision-making process.

Another big issue has to do with conflict of interest. Physicians are sometimes in circumstances where they have a vested interest in a procedure. In some cases, it can be as simple as income: one procedure is more profitable—not necessarily more effective—than another. Physicians also often have relationships with vendors. For example, a surgeon might have been instrumental in helping a company create a particular device, so he wants hospitals to buy it. Or a physician may have recently accepted a paid speaking engagement to address a particular pharmaceutical company. This type of problem has become prevalent enough that some academic institutions are banning pharmaceutical industry–sponsored gifts, food, and trips for physicians who affiliate with these universities.

This chapter presents the ethical issues that have confronted medical professionals in recent years. These issues range from Baby K to the more recent pillow angel case. The chapter presents questions but few answers. There are no

This cartoon depicts the dilemma of the physician who is caught in a conflict between good medical practices and what is permitted by the insurance company. *(Jeff Sebo)*

easy ways to manage many of these issues—well-meaning people and perhaps an organized hospital-based system for making decisions may be the best way to wade through some of the problems that are guaranteed to present themselves.

BABY K

On October 13, 1992, the infant known as Baby K was born in Fairfax, Virginia, without a brain—a condition referred to as *anencephalic.* The brain stem, the portion of the brain that regulates the heartbeat, blood pressure, and respiration was intact, but there was nothing else there. The Fairfax Hospital doctors were in favor of a do not resuscitate order for the child, but the mother wanted the hospital to provide advanced supportive care (which primarily consisted of ventilator support). In the opinion of the medical professionals, treatment was futile.

All other hospitals refused to take the baby, and Fairfax Hospital kept the baby on the ventilator for six weeks, gradually weaning her from it. The mother then moved the child to a nursing facility, but the infant was brought back to the hospital many times for respiratory difficulty. At the age of six months, the baby was again admitted to the hospital, and at that point, the hospital filed a legal motion for a guardian to be appointed by the court. They sought permission from the court for the doctors to provide only *palliative* care, which would not have included the ventilator. The case went to trial and, while the mother argued for sanctity of life, the hospital experts argued that ventilator care for a baby without a brain went beyond the accepted standard of medical care.

When the ruling came out from the U.S. District Court for the Eastern District of Virginia, the controversial ruling called for the hospital to put the baby on a mechanical ventilator whenever Baby K had trouble breathing. The court's ruling stated that in its interpretation of the Emergency Medical Treatment and Active Labor Act (EMTALA) patients who present with an emergency must be stabilized. The court refused to address the moral or ethical issues; they said they just wanted to interpret the laws as they existed. Baby K lived almost three years, until 1995.

The case raises many interesting issues regarding bioethics. The medical community is in agreement concerning the irreversibility of anencephaly. Many point out that to drain the energy of medical personnel and to spend money on the support resources for a baby with no hope is not the purpose of well-advised medical care. Some doctors also point out that consulting an ethics board interferes with a medical doctor's right to make a sound medical decision.

KAREN ANN QUINLAN

The case of Karen Ann Quinlan (1954–85) is equally heartbreaking, and it highlights the right to die controversy in the United States. In 1975 at the age of 21, Quinlan went to a party where she consumed Valium and alcohol. Once home, she collapsed and stopped breathing twice for 15 minutes or more. When the paramedics arrived, they took her to the hospital and she slipped into what is called a persistent vegetative state. (Her brain showed only abnormal slow-wave activity.)

She was kept alive on a ventilator for several months and had to be fed via a feeding tube. When there was no change, her parents requested that the use of the ventilator and other forms of active care end and she be allowed to die. The hospital refused, and legal battles and a great deal of publicity followed. Eventually the court ruled in her parents' favor and she was removed from life support in 1976. However, she did not die. For nine years she was fed through the feeding tube, and she existed in a coma for almost a decade, finally dying of pneumonia in 1985. Quinlan's father was assigned legal guardianship for his daughter, and the family was devoutly Catholic. The fact that the court found in their favor may have partially stemmed from the fact that the Catholic Church's moral teaching of basic care but not extraordinary means was factored in by both sides.

The case spurred significant changes in the practice of law and medicine around the world. Hospitals determined that it was important to create formal ethics committees to advise or be involved in these types of issues, and attorneys and health care administrators began advocating for health directives. Today, people make living wills that stipulate how they would like end-of-life issues handled if they are not conscious and able to make decisions for themselves.

THE PILLOW ANGEL CASE

In 2007, the pillow angel case received a lot of press attention, partly because of the ethical dilemmas it introduced. Ashley is a 9-year-old child who was born with a rare brain condition known as static encephalopathy; she has little control over her body and is dependent on her parents for all of her care. Her parents feared that as she got bigger it would be harder for them to take care of her, so she has received high-dose estrogen treatments (the type used when gangly teen girls do not want to grow too tall) to stunt her growth, and they have removed her uterus to prevent menstruation and her breast tissue because of a family history of cancer and fibrocystic disease.

As a result of these treatments, Ashley will remain at her present height of 4′ 6″ (1.4 m) and weight of about 70 pounds (32 kg). Without the procedures she would reach about 5′ 6″ (1.7 m) and weigh about 120 pounds (54 kg). In their blog, her parents give two major reasons for the procedure: It will be good for Ashley, and it will be good for the parents.

The family lives in Washington State, and while her parents were very certain it was the right thing to do, the decision was more complicated for her physicians, who also called in a representative of the bioethics committee of the American Academy of Pediatrics. They had several worries about side effects. They had to be sure there would be no medical harm. The estrogen treatment bore the risk of blood clots, but for the most part that is not common among younger people. Removing breast buds is a much less invasive procedure than a mastectomy. The ethics committee essentially did a cost-benefit analysis and concluded that the rewards outweighed the risks. Keeping Ashley smaller and more portable, the doctors argued, has medical as well as emotional benefits: more movement means better circulation, digestion, and muscle condition, and fewer sores and infections.

The ethics committee gave weight to the fact that the parents sincerely loved their daughter and saw this as a solution that would permit them to keep her in their care at home. (Home care becomes more difficult as people get bigger simply because it is harder for a layperson to manage the equipment and the difficulty of getting a bigger person around.) But disability rights advocates took issue with the reasoning that it was okay because it was "out of love."

"Benevolence and good intentions have been among the biggest enemies of disabled people over the course of history," says Arlene Mayerson, a leading expert in disability rights law, who was quoted in an article on the topic in *Time* magazine (January 22, 2007). "Many things that were done under a theory of benevolence were later seen as wrongheaded violations of human rights." With the right information and support, disability rights advocates believe, there is no need for a medical solution to an essentially social problem.

THE ETHICS OF MEDICAL STUDIES

By the late 20th century, researchers were beginning to realize that studies on various diseases or treatments were sometimes not coming out as they expected; they realized that a clinical study consisting of mostly men, for example, provided information that often did not apply to women. As a matter of fact, the information was often shockingly different.

In the mid-20th century, Dr. Lawrence Craven, a physician in California, noted that children who were given aspirin after having their tonsils removed often had bleeding problems. He began to think that it was the aspirin product that prevented the blood from clotting, and it occurred to him that perhaps this would help with heart disease. After examining the medical records of 8,000 aspirin users, he found that none had had heart attacks, so he invited other physicians to examine the issue.

In 1983, Dr. Charles Hennekens of Harvard Medical School undertook a study of 22,000 healthy male doctors, all of whom were over the age of 40. The study showed a reduced rate of heart attack but a greater likelihood of bleeding in the brain. This information has led to the recommendation that people with bleeding problems should avoid aspirin, but that other people should take a low dose to maintain heart health.

Finally in 2005, a study of women was conducted. The results of this study showed that for women aspirin did not reduce the risk of a first heart attack but it brought down the risk of stroke by 17 percent. These results were counter to what was learned about men.

From studies such as these, scientists are now aware that gender is one more variable that needs to be carefully factored in when diagnostic methods and treatment remedies are being studied.

CORRECTING ETHICAL MISTAKES OF THE PAST

The ethics of a flawed study from 1932 were not resolved until 1972 when a public health administrator, who had been angling for six years to bring attention to the issues involved in the study, finally leaked what he knew to the press.

In 1932, the U.S. Public Health Service started a medical evaluation that was based on a study used in Oslo that patched together information about the course of untreated syphilis. At the time, standard medical treatments for syphilis were widely known to be toxic, dangerous, and not necessarily effective. The idea behind observing syphilis untreated was twofold: to determine if patients did better without the toxic cures, and to identify the stages of syphilis with the idea that a stage-specific treatment might be effective. To buttress what the Norwegian researchers were doing with a retrospective study, U.S. scientists chose a county—Macon

County in Alabama—with a very high rate of syphilis and also a high rate of poorly educated African Americans. Six hundred poor black men were put into the study conducted by the Tuskegee Institute and the Veterans Administration with promises of free medicines, regular medical care, burial assistance, free hot meals on the days of examination, transport to and from the hospital, and sometimes an opportunity to shop while in town. Of the group, 399 were thought to have syphilis and 201 were the control group.

The study group was frequently misled as to what was happening to them; treatments were often nothing more than *placebos,* and spinal taps for evaluative diagnosis were billed as "Last Chance for Special Free Treatment." Then in 1947, another boundary was crossed when penicillin began being used effectively against syphilis. The decision was made to continue the study and also to continue to withhold penicillin from the men without telling them that an effective treatment had been found.

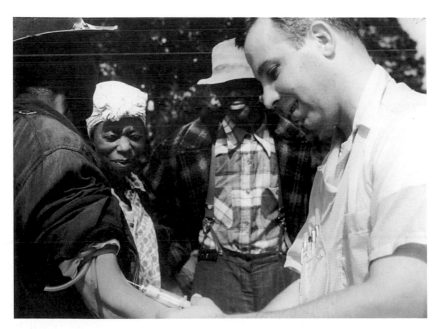

Unidentified subject and onlookers as Dr. Walter Edmondson conducts a blood test as part of the study *(National Archives and Records Administration)*

In 1966, Peter Buxtun, a Public Health Service investigator, filed an official protest with the Division of Venereal Diseases of the Centers for Disease Control. His objection was ignored since the study was not yet complete. (The study was to be considered complete when all participants had died.) He raised the issue again in 1968 and was ignored. At risk of losing his job, he finally decided his only option was to leak the story to the *Washington Star,* which he did in 1972. The reporter Jean Heller's story appeared in the paper on July 25. Other newspapers picked up on it, and the study was quickly brought to an end in November 1972 when the press turned public sentiment against the methodology. By this time, 28 men had died of syphilis, another 100 had died of complications related to syphilis, at least 40 wives had been infected, and 19 children had contracted the disease at birth.

If ethics boards had been in existence in the 1930s, would this study even have been permitted? Ultimately, the participants won an out-of-court settlement; each survivor received $37,500 in damages and the heirs of the deceased received $15,000. Since that time, the government has created a method to evaluate its research practices and to monitor all studies using human subjects.

THE CREATION OF HOSPITAL ETHICS BOARDS

The field of medical ethics is very young, and the first interest in creating ethics committees within hospitals was in 1971 when the Catholic Hospital Association of Canada recommended that Catholic institutions educate the hospital community on how to weigh the use of life-sustaining technologies. From that beginning, hospitals throughout North America began implementing ethics committees to help evaluate many of the difficult decisions that physicians were encountering in health care.

Some committees focus primarily on hospital policy or implementing educational panels so that staff members are better prepared to make their own decisions. Other times, these committees are asked by patients, families, nurses, or doctors to recommend action—or inaction—on difficult ethical questions. These

VALUES IN MEDICAL ETHICS

Committees addressing issues of medical ethics always intend to take into consideration the following:

- Non-*maleficence.* First do no harm (*primum non nocere*).
- *Beneficence.* A practitioner should act in the best interest of the patient.
- Patient autonomy. The patient has the right to refuse or choose their treatment.
- Justice. If health resources are scarce, then there needs to be justice in deciding who gets what treatment. This issue could pertain to anything from situations such as the lack of resources in New Orleans during Hurricane Katrina to the ongoing issues of a shortage of organs for transplant.
- Dignity. Both the patient and the person treating the patient have the right to dignity.
- Truthfulness and honesty. Part of this concept involves informed consent, and this concept has become increasingly important over time. People need to be given a clear description of what is before them in order that they can make an informed decision about going forward. (Until the 1970s, American culture did not emphasize truth-telling in cases of cancer. Today, patients are asked if they want to know the full diagnosis.)

The ability to legislate or regulate ethical judgment is simply impossible. Every case presented varies, and the best hospitals or medical groups can do is work to establish a sensitive, intelligent, ethical group of people who can

(continues)

(continued)

discuss each case rationally and carefully. With a little luck, these groups will make the best decisions possible given the information provided them.

Patients themselves often are placed in a difficult predicament with not enough information to make a very emotional decision. In this case, the best path may be clear communication. Patients need to be very clear about their worries as well as their hopes for the outcome. A patient advocate (often a friend or family member) can be instrumental in helping the patient make decisions. Even if the patient basically feels well enough to manage his or her own care, it is often difficult to absorb all the information being presented, and having another person who is part of the process can make it less stressful for the patient.

questions range from how long and whether to extend life using new technology, how things intersect with religious beliefs, how, whether, and when to consider extraordinary costs as a factor, and how much to respond to consumer demand. None of these questions are easy. Boards may also encounter very specific cases: Which patient should have an organ transplant? Who will be taken off life support? Should someone near death be resuscitated? How aggressively should new technologies be employed?

Members of ethics boards are generally appointed by the hospital and are usually hospital administration or medical staff. A member of the public and a chaplain also may be included. Some hospitals are adding experts in bioethics. Many offer training for the committees to help members in their decision-making.

It is new territory, and not all doctors embrace the proffered service. Some want to be able to make the decisions themselves; others say there often is not time to consult an outside group; still others fear that the process will make a difficult situation even worse.

At this point, there are no studies or statistics that provide helpful information on whether these committees are a plus or minus to medical care. Over time, these entities will almost certainly be reshaped and refined to provide helpful guidance for difficult issues. One doctor pointed out that sometimes it is simply helpful to have a group to use as a sounding board; by talking it through and getting suggestions from other people, sometimes the physician can then see the situation with more clarity, which can either bring about a better answer or provide suggestions as to how best to help loved ones who face a difficult circumstance.

CONCLUSION

There is no doubt that the medical community faces a growing number of difficult questions. As more technology permits more life-saving options, the results do not always guarantee an optimum way of life. Medical ethics boards are one possibility for helping society navigate treacherous waters. Another might simply be common sense: Clear communication from both patient and physician can often resolve many potential conflicts.

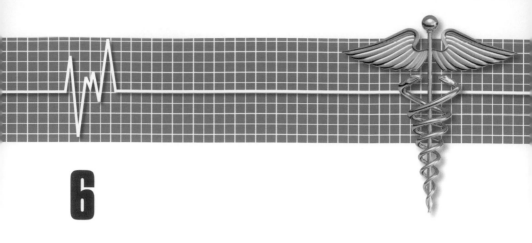

6

Health Care in Crisis—
Who Will Pay?

A merica is known for its generosity in sending emergency responders to help with medical issues around the world. Major philanthropic organizations such as the Red Cross and Médecins Sans Frontières (Doctors without Borders) provide helpful gear such as mosquito nets and send in food and medical care to disaster sites. Despite this active approach to taking care of the world's inhabitants, the health care picture at home is far from solved. In the United States in the first decade of the 21st century, 47 million Americans lack health insurance, meaning that easy and affordable access to health care is out of reach for an astonishing number of Americans. This may mean very little to the healthy 24-year-old who can not envision a day when he might need health care, but it may make a big difference to a person with debilitating arthritis who needs to be on maintenance medications.

Since 2000, the number of uninsured has increased by more than 20 percent, reaching 47 million in 2006. The numbers of the uninsured have grown in the 21st century as the rates of employer-based coverage have stalled or declined. While there has been an increase in the number of children who are covered by the Medicaid State Children's Health Insurance Program (SCHIP), over the last 10 years, it does not fully compensate for the loss of job-based coverage for other Americans. Young adults, ages 19 to 29, have the highest uninsured rate (30 percent) of any age group. More than half of these people are full-time workers, but their low incomes make it difficult to afford coverage. Members of minority groups are more likely to be uninsured.

In addition, rising health care costs are squeezing many middle-income Americans, even those with insurance, who report difficulties paying medical bills due to a lack of adequate coverage. The Kaiser Foundation states that 25 million people are underinsured. These people are more likely not to seek care because they know there will be very high out-of-pocket costs.

To add to this negative news, Americans have the most expensive health care system in the world, but we have no bragging rights over outcomes. The United States ranks 29th in infant mortality, 48th in life expectancy, and 19th in preventable deaths. These are shocking statistics for a country that is supposed to be one of the most socially advanced in the world.

Not having health insurance is a problem that is much bigger than what would already be the daunting worry of how a family is going to pay medical bills. It affects people's access to medical care, since preventive care drops by the wayside. In many cases, this snowballs into a personal and a national disaster; the diabetic who could have kept his disease under control if he had had access to preventive advice and the necessary medications eventually finds himself in the emergency room—the medical bill eventually being absorbed by the government. (The cost of care for the uninsured is paid through surcharges to private payers, which results in higher private insurance premiums; the remainder is absorbed by the government.) The patient may face anything from a few days in the hospital to an amputation, an unfortunate and serious problem with untreated diabetes. In the long run, this costs society.

There are two other major problems with the system we have today. The first has to do with the burdensome workload placed on every physician who accepts insurance. It is estimated that today's physician spends about one-third of his time satisfying insurance

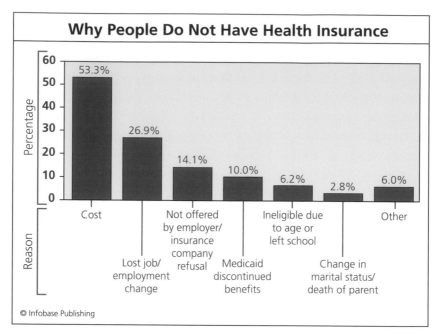

Why People Do Not Have Health Insurance

© Infobase Publishing

The number of Americans with no health insurance is a major problem in the United States at the beginning of the 21st century. This illustration indicates the major reasons.

company regulations and seeking approval for treatment, wasting time that could be better spent with patients.

In addition, the fragmented system used in this country has made it impossible to rein in escalating costs. One reason costs keep going up has to do with the complexity of administering such diverse programs. An estimated 10 to 40 percent of health insurance premiums go toward the cost of running the program, not toward health care. This chapter will explain health insurance programs in America today, offer a peek at what other industrialized countries offer, and present some of the options that are being considered to remedy the situation.

THE BACKGROUND ON HEALTH INSURANCE

Until the mid-19th century, no Americans had any form of health insurance. If a person got sick and could afford to go to the doctor, they paid directly for services rendered. Starting in the 1860s, insurance companies started offering plans to insure against accidents from travel by rail or steamboat; perhaps the opportunity to make some money on people's fears was just too good to pass up. However, the process of selling health insurance opened the door for selling more comprehensive health plans covering other types of illnesses and injuries.

Group health insurance plans got a boost with a program in Dallas, Texas. In 1929, a group of teachers made a deal with Baylor Hospital for room, board, and medical services in exchange for a monthly fee. Shortly after this, other insurance companies began offering health policies, the best known of which was the Blue Cross and Blue Shield plan that was first offered in 1932. Blue Cross and Blue Shield was a nonprofit organization that grouped people who needed services. They promised volume business and prompt payment to physicians and hospitals and, in return, the organization received discounted health rates.

In the 1940s, an increasing number of companies offered health plans. The government imposed wage freezes because of World War II (1939–45), and many companies also froze wages. They found that offering health care benefits permitted them to win over the more desirable workers. In the 1950s, unions bargained for better benefit packages and the process grew.

In the late 1950s and 1960s, the government began to expand its programs to cover health care costs. In 1954, disability benefits (payment for workers who become disabled and are no longer able to work) were added to social security coverage for the first time. Then in 1965 the government created Medicare and Medicaid (described later in the chapter).

During the last two decades of the 20th century, the cost of health care in America grew quickly, and this has forced companies that were facing ever-increasing insurance premiums to figure out different programs to offer their employees. In the process, companies have been pushing their employees into more affordable health maintenance organizations or in some cases dropping the health program altogether. By 1995, individuals and companies only paid for about half of health care; the government took responsibility for the other half.

In 1993, President Bill Clinton pushed Congress to reform the laws so that all Americans would be guaranteed health insurance, but Congress opposed the plan. The majority of members felt it was too expensive. Alternatives were offered by various members of Congress, but nothing was passed. One step forward was taken in 1996, when Congress passed the Health Insurance Portability and Accountability Act. Reflecting the increase in job mobility for Americans today, this provision protects individuals from losing their health insurance when they move from one job to another or become self-employed. However, the coverage is expensive and its availability is limited, so it provides no long-term solution.

President Barack Obama has entered his term of office with promises to reform health care. While full-scale revamping of the program is unlikely, there are numerous options that the president will likely push for as he continues in office (discussed later in the chapter).

THE MEDICAID PROGRAM

Medicaid is the nation's public health insurance program for low-income Americans, covering 59 million low-income children, families, seniors, and people with disabilities. The program is a federal-state partnership that provides coverage based on income levels and eligibility categories. Medicaid beneficiaries are much poorer and in markedly worse health than the privately insured population. Medicaid is designed as a safety net during difficult economic times, but when states encounter budget shortfalls as enrollees increase, it becomes difficult. About 14 percent of the nonelderly population is covered by Medicaid, making it the largest provider of health insurance.

Medicaid guidelines specify four main categories of the nonelderly low-income individuals that it will cover: children, their parents, pregnant women, and individuals with disabilities, but this still leaves 35 percent of those below the poverty line with no insurance. Over the past decade, growth in Medicaid enrollment has helped buffer losses in job-based coverage.

Medicaid and SCHIP cover more than one-quarter of all children and a little more than half of low-income children. SCHIP was created in 1997 to expand coverage for children who were caught in a middle ground (and the 2009 federal budget provided for further expansion of the program). Their families were not desperate enough for Medicaid, but they could not afford basic health care without government help.

While Medicaid covers the bulk of low-income children in the United States, SCHIP is designed to supplement Medicaid by covering children whose family income is too high to qualify for Medicaid. Between Medicaid and SCHIP, half of all low-income families are covered. But lack of understanding of the programs means that many who are eligible are still not taking advantage of them.

Medicaid for those with disabilities is limited to those with very low incomes and few assets but is available for people with conditions like HIV/AIDS.

Medicare is a federal program run by the U.S. Social Security Administration that reimburses hospitals and physicians for medical care provided to qualifying people older than 65.

HOW MOST AMERICANS RECEIVE HEALTH INSURANCE

Employer-based coverage as a job benefit takes care of more than half (61 percent) of Americans under age 65. Most of those over the age of 65 qualify for Medicare, and SCHIP covers children who do not have health insurance. But the gaps between public and private health care today currently leave 47 million Americans without health insurance.

Employer-sponsored health insurance is voluntary; businesses are not legally required to offer it and employees can opt out of it. Part-time employees and the newly hired generally do not have coverage. If a person works at a company that does not provide health care, it can be difficult or very expensive to buy an individual health policy. The person with a chronic illness, someone who has had a heart attack, or someone with a disability can find it very difficult to find insurance; insurance companies try to exclude people who might have high medical needs as that will cost the insurance company more money than it would cost to insure a healthy individual. According to the Kaiser Foundation, in 2005 nearly three out of five adults who considered buying coverage had difficulty finding a plan they could afford, and one in five were denied coverage, charged a higher price based on their health status, or had a specific condition excluded from coverage.

While there are federal incentives in the form of tax breaks for both employees and employers who participate in health insurance programs, the fact that there are still 47 million Americans without insurance shows that they are not enough.

THE PROBLEM WITH HAVING SO MANY UNINSURED

A lack of insurance affects a person's health. Studies reveal that one-quarter of uninsured Americans skip medical visits because they are worried about paying for the care. And even if they seek treatment, it is often too costly to pursue. Medicines are costly, so prescriptions go unfilled. The uninsured have to evaluate a visit to the doctor against paying the rent or buying food. One-quarter of all adults, according to the Kaiser Family Foundation (2008), report that at some point in the past five years they spent less on their basic needs in order to pay medical bills.

Hospitals discourage being used as doctors' offices, so those who come in are often charged market rate as opposed to what health insurers or public programs would pay for the same service. Care for the uninsured is generally left to hospitals, community clinics, and some physicians. When health care services are provided to those who cannot pay for them, a patchwork of federal, state, and private funds have to pick up the cost. The bulk is funded by the government so in the long run, all taxpayers pay. The cost of care for the uninsured amounted to about $57 billion in 2008.

POSSIBLE SOLUTIONS TO THE AMERICAN DILEMMA

Public interest in finding a solution has risen during recent years, though there is little public or political accord as to how to achieve this goal. Although there are a variety of ideas for organizing the health system and different ways to provide coverage, in general all the plans share the common goal of expanding coverage so fewer people have no insurance and also to subsidize coverage for the poor.

While the idea of universal health care tends to make people nervous, all universal health care really means is giving quality coverage to every American. The following are some of the ways the United States might achieve this goal.

1. Build on the current system
   a. Some experts favor expanding employer-based coverage and public insurance programs and redesigning existing private markets so that people can obtain coverage outside the employer system. Right now three-fifths of the nonelderly have employer-based coverage. The government could increase this by requiring large employers to offer coverage, with incentives offered to small employers to voluntarily offer health benefits. Many also favor the play or pay proposals, which require employers to provide health coverage (play) or pay into a pool to help finance the cost of coverage for their employees. To ease the economic burden on employers, governments could offer tax credits or premium subsidies.
   b. Build on Medicaid and SCHIP by expanding them.
   c. Offer new group insurance options. For those without access to employer-based coverage and who are not eligible for public programs, some proposals would create new national or regional pools where plans could be offered to individuals or employees.

Premiums could be subsidized for those who could not afford them.

2. Change the tax system

The tax code could be changed to offer better incentives. Since employees change employers often now, the job-based coverage may be outdated. The country needs a system that can be maintained across jobs and state lines.

HOW MASSACHUSETTS PUTS ITSELF ON THE MAP

In 2006, Massachusetts became the first state to enact a near-universal health care reform plan. Within 18 months, nearly 440,000 people had gained coverage, cutting the state's uninsured rate in half. The mission of the 2006 Massachusetts health plan was clear: Every resident had to be covered. A board known as the Health Connector was established to match individuals to the right plan for them, and subsidies were given to individuals who could not afford insurance. Some of the measures implemented in Massachusetts involved:

- Expansion of Medicaid. By making more of the poor residents eligible for the state program, they were able to expand who was covered.
- New aid for the working poor. The state created a new program to subsidize coverage for low-income individuals.
- They established private plans that would sell insurance to people who earn too much for the basic program for the working poor.
- New options for young adults (19–26). Most young adults, particularly males, consider themselves healthy and not in need of investing anything at

3. Establish a single-payer system.

Many feel that the inefficiencies and gaps in the system are only remediable by switching to a single-payer system. This would mean all health care coverage would be managed and paid for by the government. One proposal would be for a national insurance plan modeled on Medicare that would contract directly with private

all in health insurance. By bringing the cost down for these individuals, the program aspires to better suit their needs.

Under the Massachusetts system, companies with 11 or more workers that do not offer insurance to their employees are required to help finance the public program on a per employee basis. The state is also adding money from the Medicaid program out of savings that it normally would have paid to hospitals and clinics each year for care of the uninsured. A benefit to this program is that it can expand as job-related insurance may drop.

Making health insurance mandatory has presented its challenges. The state has spent the past year dealing with questions about how much basic coverage people need and how much they can be expected to pay. But cost continues to be a problem. This year Massachusetts is exempting 62,000 people from the statewide mandate because they cannot afford insurance even with government subsidies.

Many other states have expanded coverage, but with the economic downturn the ability for states to expand this system is hampered. However, as the cost of health care rises, there is going to be increasing need to focus on this issue.

providers and possibly insurance companies. Another proposal is to create a new government-financed health care system in which individuals purchase insurance on their own through state or regional purchasing pools.

As Congress wrestles with what options to put into the health reform act that President Barack Obama has put on his priorities list, there is one aspect on which all agree: There is serious need for some type of reform. Time will tell how this all plays out.

HOW OTHER COUNTRIES MANAGE

At $2.3 billion, the United States pays twice as much in health care dollars per capita as any other nation, and the cost is rising. Insurance premiums have jumped 78 percent during the past eight years while wages have gained only 15 percent, and the United States spends 15.3 percent of its gross domestic product (GDP) on health care, more than any other country. It is instructive to consider how other countries manage.

- The United Kingdom is either hailed or reviled for what is described by some as socialized medicine, meaning that the government both provides and pays for health care. Britons pay taxes for their health care, and the government distributes those funds to the National Health Service, which then pays the health care providers. Hospital doctors are salaried, and general practitioners who run private practices are paid based on the number of patients they see. The administrative costs are low; there are no bills to collect or claims to review. Patients have a general practitioner who helps them navigate the system, and these physicians are rewarded for keeping their patients healthy. (Britain is the world leader in preventive health care.)

  The downside to the program generally involves long waits for nonemergency care. If a person needs knee replacement surgery, they will get it, but they are generally on a waiting list until there is time. Britain is modifying the system to try to improve it, but generally it is an available and helpful program. (Those who opt to seek private care for certain services can do so if they are willing to pay.)

■ Japan uses a system of social insurance where all citizens must have health insurance, either through their work or through a nonprofit plan. Public assistance is given to those who cannot afford the premiums.

■ Germany works with a system similar to Japan's. Germans can buy their health insurance from one of more than 200 private nonprofit sickness funds. These funds are not allowed to deny coverage based on preexisting conditions, and they compete for having the largest enrollments to keep it competitive.

■ Taiwan adopted national health insurance in 1995 and modeled it after Japan and Germany, but there is only one government-run insurer. Taiwan also implemented the use of a smart card that stores all of a person's medical history on it. This improves care.

■ Switzerland uses social insurance, but in 1994 when they passed the bill 95 percent of the population already had coverage. Switzerland is often discussed as an example of a capitalist nation where the citizens have full coverage. Though there are powerful insurance and pharmaceutical businesses, they are not allowed to cherry-pick the healthy and not cover the ill. While it is the second-most expensive system in the world, it is still cheaper and more effective than what exists in the United States.

There is no doubt that America could learn from the experience of other countries. While other programs work better, they are not without flaws. For example, neither Taiwan nor Japan have yet worked out a system to collect as much as they need, so the programs are currently run at a deficit.

WHY ELECTRONIC HEALTH RECORDS ARE IMPORTANT

Today, a person can sit down at a computer and find out a great deal about almost anything—from finding a video about home plumbing repair to learning where the neighbors contributed politically. The one thing Americans do not have access to on any computer is their own health records. Eight out of 10 physicians still work only with paper records, either making notes on paper that is thrown in a file, or perhaps dictating notes into the telephone where the patient information is turned into a document, but not one that is accessible to anyone else. If a patient goes to a facility for a mammogram, an X-ray, or a scan of some type, that record is now stored digitally, but these records are often stored in a form that is unique to that facility and not available to others.

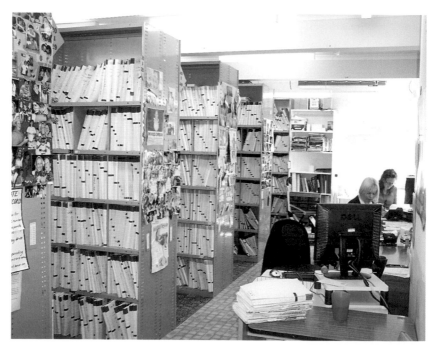

Until all medical records are electronic, this is how they continue to be stored. *(Echuca Regional Health)*

To the patient with any sort of complex problem, this creates a tremendous dilemma. A very simple example might concern a person with back pain; he may have consulted a physician who recorded notes on the visit by hand and then sent the patient off to get a scan of the spine. The scan was conducted, evaluated, and sent to the physician who ordered the test. In the meantime, some of the exercises and the pain reliever prescribed by the physician have begun to work, and the patient forgets about the problem. Then he goes on a ski vacation where he reinjures his back. Doctors in the ski community where he seeks help have to start all over again. It is a Saturday, the patient's records are in handwriting in his own physician's office 200 miles away, and the scan of his back is digital but the patient has no way to get that scan sent to the doctor treating him that day.

Imagine how different—and how much less costly—it would be if hospitals, medical testing facilities, and physicians were required to maintain electronic records (in a format that could be accessed by all facilities in the United States). Or patients could add a medical records USB flash drive to their key chain so that if and when they ever needed access to test results or past medical history, they could simply provide the data storage unit to the physician whom they were consulting.

Many people worry about privacy. They fear that if patient health records were electronic, hackers could find ways to access information and use that information—anything from a former drinking problem to a chronic illness—to embarrass a person or discriminate against them. Most medical professionals feel there is no reason to worry about privacy of health records. The 1996 Health Insurance Portability and Accountability Act ensured that medical records could not be accessed without permission by employers or anyone else without medical need. While health records are available to insurers and could cause problems, for the most part there is no reason to worry.

Information technology for health care received $20 billion of the $787 billion stimulus package in 2009, so it is hoped that hospitals and physicians will set up systems that provide an over-all network to give access to anyone involved in providing medical care. This would improve the quality of care patients receive.

CONCLUSION

While progress on electronic records may inch along over the next few years, there are still many challenges to overcome before major health care legislation becomes a reality. This topic is of high importance. All citizens and government leaders need to study what the possibilities are and advocate for what seems the most workable. Affordable health care could improve the medical care of young and old—the very least a wealthy country should provide its citizenry.

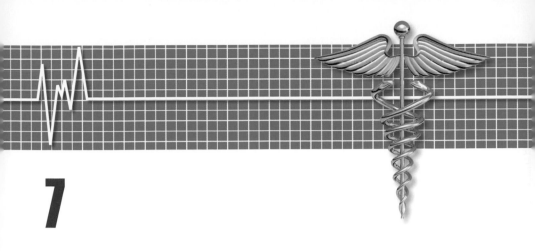

7

Preventive Medicine— Key to Better Health

Anyone who picks up a newspaper or surfs the Internet will find stories of strange illnesses and amazing cures. While these illnesses make headlines, they are not actually the illnesses that kill people in large numbers. As charts in this chapter reveal, the leading causes of death in the United States are the same old diseases that have been around for the last century: heart disease, cancer, and stroke are by far the leading causes of death. Also instructive is a look at the chart that depicts the underlying causes of death. These too are nothing new. Tobacco use, poor diet, lack of exercise, and alcohol consumption, all self-inflicted, are the major causes of illnesses that bring about the highest numbers of death.

Another interesting fact is that very few people stay home sick with unique or difficult-to-diagnose illnesses. People miss school or work because they have colds or the flu. These illnesses sound boring, but anyone who has suffered either recently knows that they can cause several days of real misery.

Society also faces another problem: food safety. Few people grow their own food now, so people must rely on the safety of the food that is delivered to their stores. As recent news headlines have proven, this is not an easy goal.

While it may not seem exciting, this chapter stresses how important it is that medical professionals, governments, and individuals pay attention to these highly preventable factors that can lead to chronic illness and death. The epidemic of obesity will be addressed, as will what is new in colds and flu. Food safety will also be highlighted. Another growing problem in our increasingly sophisticated society is an increase in autoimmune disorders.

THE NEW EPIDEMIC—OBESITY

The annual deaths in the United States attributable to poor diet and inactivity hover at 300,000, and one in three, or 58 million American adults aged 20 through 74, are overweight. According to data from the Third National Health and Nutrition Examina-

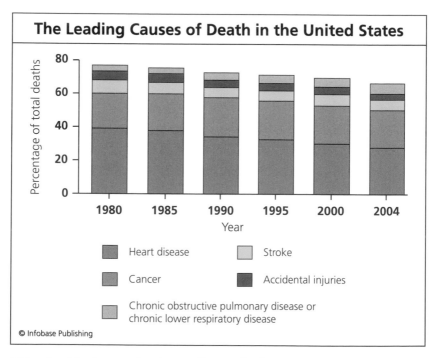

This chart indicates that heart disease is the leading cause of death in the United States.

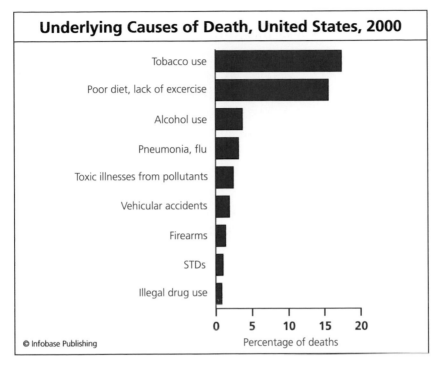

Underlying Causes of Death, United States, 2000

Tobacco use

Poor diet, lack of excercise

Alcohol use

Pneumonia, flu

Toxic illnesses from pollutants

Vehicular accidents

Firearms

STDs

Illegal drug use

0   5   10   15   20

Percentage of deaths

© Infobase Publishing

When scientists investigate the reasons behind the leading causes of death, this chart reveals what they find.

tion Survey (NHANES III), this number is growing. Minority populations are particularly affected. Other statistics show that more than one in five children and adolescents (6–17) are also overweight.

These statistics predict major health problems for Americans. Experts say that these statistics mean that this generation of young people may be the first generation who is not expected to outlive their parents. Being overweight is a known risk factor for diabetes, heart disease, high blood pressure, gallbladder disease, arthritis, breathing problems, and some forms of cancer. Extra weight places extra stress on the heart (nearly 70 percent of the diagnosed cases of cardiovascular disease are related to obesity) and more than doubles one's chances of developing high blood pressure, which affects approximately 26 percent of obese American

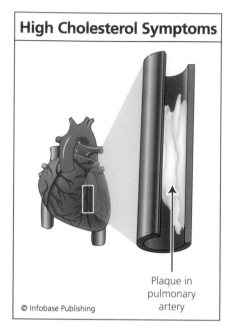

High Cholesterol Symptoms

Plaque in pulmonary artery

© Infobase Publishing

Physicians know high cholesterol causes heart disease, but convincing people to eat right is not easy.

men and women. However, it is surprising that cancer and obesity seem linked. Almost half of breast cancer cases are diagnosed among obese women; an estimated 42 percent of colon cancer cases are diagnosed among obese individuals. Obesity-related breast cancer and colon cancer account for 2.5 percent of the total costs of cancer, or $1.9 billion annually.

In the United States, the Centers for Disease Control and Prevention (CDC), through their public health information offices, is pushing toward a leaner public. Knowing that all solutions must ultimately be local, the CDC has initiated several partnerships to try to bring home the message about healthier eating.

- The Healthy Eating Active Living Convergence Partnership (CP). One problem for the poor is lack of affordable alternatives to promote healthy eating. The CP and the CDC work to encourage more active lifestyles and improve access to healthy foods.
- Early Assessment of Programs and Policies to Prevent Childhood Obesity is designed to help communities evaluate programs that help teach low-income children to improve eating habits and physical activity levels.
- Addressing Obesity through Commercial Health Plans. With this program, the CDC is working to help public health professionals and health care plan administrators

collaborate to improve obesity interventions designed for medical settings.

Other education programs are coming from other sources. First lady of New York State, Michelle Paige Paterson, has an advanced degree in health services management and has worked within both hospitals and for health care providers. She has resolved that combating childhood obesity would be the primary focus of her work as first lady. She has launched "Healthy Kids, Healthy New York," a six-week program for middle school students to encourage them to track their eating and exercise habits.

"Childhood obesity is an epidemic that is threatening our families' long-term health," Paterson said in a press statement. "Despite the incredible medical breakthroughs of the past few decades, our children are part of the first generation of kids who may have a shorter life expectancy than their parents. We have seen an increase in the rates of obesity-related illnesses, including diabetes and cardiovascular disease. We, as parents, educators and

Biking and other forms of exercise are an important part of health maintenance. *(Alabama Bureau of Tourism and Travel)*

Farmers market—a reliable source for fresh food *(Katie Mermaid)*

community leaders, must promote healthy activity, eating and living so that our children can lead long, healthy and active lives.

"Exercise and nutrition not only improve physical health, but also affect how our children perform in school," the first lady added. "Poor nutritional status and hunger interfere with cognitive function and are associated with lower academic achievement. Healthy students are four times more likely to do better in school than obese students. These problems are preventable, and must be addressed by our schools, our communities and our government."

FIGHTING SIMPLE COLDS AND THE FLU

Some diseases like the common cold or flu are baffling because they have been so difficult to prevent or cure. Colds are generally not serious, but they can make a person feel ill for up to a week's time. Flu is different—it can be deadly for the young, the elderly, or people with compromised immune systems. Each year, scientists try to predict what strain is expected to circulate in the population and develop a new vaccine against the strain. Some years, the vaccine is very effective; other years, they miss the mark, and the strain makes its way through the population.

In 2009, researchers announced that they may have successfully engineered antibodies that could protect against many strains of the virus, including the avian flu, a virus that causes considerable worry among scientists and physicians. The research project was undertaken cooperatively by Harvard Medical School, the CDC, and the Burnham Institute for Medical Research.

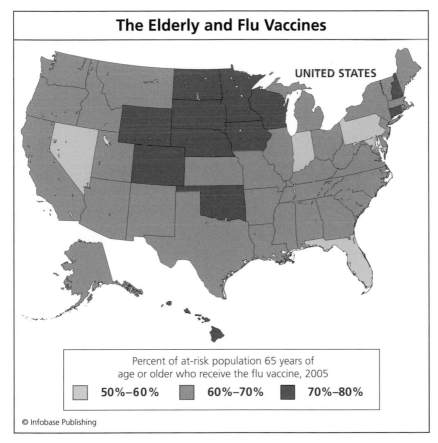

The Elderly and Flu Vaccines

UNITED STATES

Percent of at-risk population 65 years of
age or older who receive the flu vaccine, 2005

50%–60%   60%–70%   70%–80%

© Infobase Publishing

Statistics prove the importance of the elderly getting a flu shot, though not everyone understands.

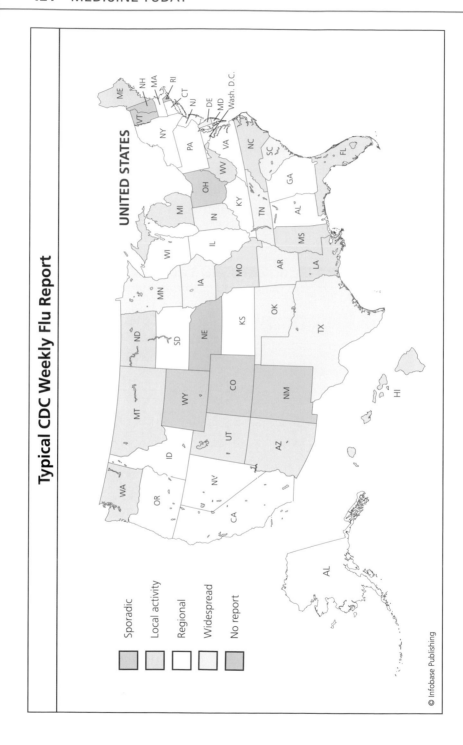

Typical CDC Weekly Flu Report

UNITED STATES

Sporadic
Local activity
Regional
Widespread
No report

Antibodies are proteins normally produced by white blood cells to attack invaders, either neutralizing them or tagging them so white blood cells can destroy them. The aspect of flu that makes it so difficult to target is that the virus has an antibody protein with spikes (outgrowths), and the spikes mutate constantly. The researchers found a way to locate the spike's neck (which they determined does not mutate), and with antibodies capable of clamping on to the neck the virus cannot mutate. While the virus can still penetrate a human cell, it cannot unfold to inject the genetic instructions that take over the cell to make more virus.

THE DANGERS OF OUR FOOD SUPPLY

According to the CDC, an estimated 5,000 people per year die and a shocking 76 million—more than double the previous estimate of 30 million—get sick from eating tainted foods. (Keep in mind that food-borne illnesses are underreported because people do not always go to their doctors for less severe stomach ills.)

There has been a recent increase in financing for food safety, partly because the government is concerned. The organisms causing illnesses are becoming more virulent, and because more food is coming from fewer sources there is the possibility for making a larger number of people ill than ever before. (Local and regional food distribution offered the advantage of limited exposure.) The government has set up eight sites around the country for active surveillance of food-borne disease, hoping they will spot outbreaks quickly and be able to stop them.

Monitoring the food supply is not simple and becomes all the more complex because there are three federal agencies involved in aspects of the process. The Food and Drug Administration (FDA), the U.S. Department of Agriculture (USDA), and the CDC moni-

---

(Opposite page) Though there are fascinating and mysterious illnesses that need to be investigated, the CDC also must find time to track basic illnesses like the flu.

tor food-related disease outbreaks. The recent national problem with salmonella-tainted peanuts that were shipped by the Georgia-based Peanut Corporation of America has brought this topic to the

---

PEOPLE RESIST SIMPLE MEASURES THAT HELP

While most people do not fear the flu, it is actually quite serious. Each year, approximately 226,000 people in the United States are hospitalized with complications from influenza, and an average of 36,000 people die—including about 100 children, according to the CDC.

Despite the fact that each year the government creates a vaccine that they hope will combat that year's strain of influenza, the vaccination rates among Americans who should be receiving them remain low. In 2009, the American Lung Association of Michigan launched a full-scale campaign to try to improve the vaccination rates. Their literature stresses the following points that should encourage businesses to urge their workers to be vaccinated:

- influenza results in billions of dollars in lost wages every year
- recent studies found that flu caused an estimated 75 million missed workdays a year
- about 200 million days of restricted activity are due to this disease that may be preventable with vaccine.

Vaccination rates remain alarmingly low, especially among working adults. In fact, adults between the ages of 18 and 49, typical ages of those in the workplace, have the lowest influenza immunization rates of any other age group.

Peanut Corporation of America headquarters *(Peanut Corporation of America)*

front pages of newspapers again. More than 500 cases of illness, including eight deaths, have been linked to products made with these peanuts. Some consumer groups are calling for the Obama administration to appoint a senior food safety official within the FDA as a better way to oversee and monitor food safety. However, the process of testing even a representative sample of the food produced in the United States—not to mention food from other countries—is a daunting process.

Pesticides and irradiation are just two of the methods that are being used in an attempt to improve the food supply. In addition, one day people may eat only genetically altered foods, hoping they are safer. All these methods are unappealing. (See the sidebar "The Solution Lies in Balance" on page 131.)

Pesticides

Part of preventing bugs from attacking spinach leaves and providing consumers with worm-free apples must be attributed to the use of pesticides. For most people, the level permitted by the federal government is safe, but in some cases the levels of toxicity on fruits and vegetables are higher than should be safe for children to consume. Children tend to eat more produce per pound of body weight than adults, and they are more sensitive to some effects of pesticides.

Eating organic foods, not exposed to chemical pesticides, offers some protection. However, nutritionists are quick to point out that organic foods should not be considered danger free. As some consumers campaign for a return to natural growing methods, the risk arises that bacteria formerly destroyed by pesticides would again invade the food supply. Ingesting bacterial microbes that may make people sick is a possibility when eating organic foods. For some families, the fact that prices are higher for organic foods is also a factor.

Irradiation

By exposing foods to ionized radiation, food scientists have discovered an effective way to destroy many of the pathogens that cause food poisoning. Irradiation has been around for 40 years, and it is likely that it will be used more and more as a method of keeping our food safe. One food scientist says: "You get a minimal amount of change in the food, no more than the nutrient destruction in cooking the food. The benefit is the food you are eating is safe."

Lester M. Crowford, D.V.M., Ph.D., director of the Georgetown Center for Food and Nutrition Policy in Washington, D.C., in a report on the safety of irradiation of food said: "It is safer to irradiate the food than to not irradiate it . . ."

Genetically Altered Foods

Scientists use two methods to alter foods genetically. The first is the traditional method, called hybridization, which has been done for years. In hybridization, plants with desirable traits are crossed to form offspring with the best traits from both parent plants.

The second and newer method involves transgenic alteration. It involves taking genes from one source and implanting them in another. By using this method, food scientists hope to one day develop broccoli with added nutrients, tomatoes with top-notch taste, low-lactose milk, and sunflower oil with less saturated fat. Food genetically altered might also deliver specifically tailored combinations of vitamins and possibly even prevent or treat ailments.

Thus far, however, most genetically modified foods are not engineered to boost nutritional content. They are supposed to boost agricultural yields by increasing resistance to insects or to weed killers, a boon to farmers who are looking for ways to grow healthier crops more rapidly.

While there is no proof that bioengineered food poses any health risks and advocates point out that insect-resistant plants will reduce the need for pesticides, there are still many unresolved issues concerning genetically altered food.

Appropriate government testing of genetically altered foods will be an important part of getting these foods to market. To date, there are no specific scientific standards for proving the environmental safety of a plant nor is there a specific review process for genetically modified foods. However, this issue and the review process are currently being studied.

Labeling is another concern. In the United States, labels identifying genetically modified foods are not required, whereas in Europe, where the debate over genetic modification of food is far more heated, labeling has recently been mandated. However, the new guidelines ran into trouble almost immediately. Additives and flavorings were initially exempt, so a food laden with additives could still claim to be without genetic modification; this is in the process of being changed. There is also ongoing discussion about how free of genetic modification something has to be to claim that status. The European Union, which is empowered to provide this type of legislation for its 27 member countries, is currently considering stipulating that products do not require labeling if each of the ingredients contains 1 percent of bioengineered material or less. Consumer groups are arguing for one-tenth of that. Unpredictable allergic reaction is another concern, though foods containing any substance known to cause allergies (such as nuts or any oil from nuts) must be labeled.

In the United States, Americans already consume food that has been genetically modified. About 55 percent of soybeans and 30 percent of corn arc modified. The new products that are being developed are different, however, because for the first time the plants are capable of reproducing. This has caused great concern among ecologists as to what this may do to the environment. If the transgene from a genetically modified plant crosses into the wild population, then a new superweed may be the result. Some genetic alterations may actually cause the stunting of growth of other types of plants with which it can be cross-pollinated.

The American public finally woke up to the alarm being expressed by Europeans when it was discovered that genetically altered corn carries a toxin that, when spread to the common milkweed plant which is a frequent intruder in cornfields, means death to the monarch butterflies who feed on milkweed. In a final twist of irony, the genetically altered plants may, in some cases, endanger varieties of plants that scientists rely on to provide the raw material for the genetic alterations.

While improved biotechnology—creating a crop that remedies the bug problem but that does not affect the plant's pollen, for example—may ultimately be an answer, there is great concern among ecologists that the U.S. government has given approval to plant these genetically changed plants without yet having seen the results of long-term studies on the cffccts on the environment. The tremendous profit potential for industry, as well as a very real pressure for increased food production, has made it tempting to adopt a short-term view of the risks and benefits of putting genetically modified crops on thc fast track.

Despite all these issues, many scientists still argue that transgenic alteration of plants is actually safer than hybridization (something that has taken place for so long no one has ever thought to question it), because working with genetic material is more precise than cross-breeding plant families. A valid point made by the naysayers is that continued testing is important to any type of experimentation done with foods.

THE SOLUTION LIES IN BALANCE

Scientists who specialize in food safety say that balance and safe preparation methods is key and stress the following points.

- People should eat a wide variety of fresh food. This is particularly helpful for children. If people eat a range of fresh foods, it minimizes their exposure to any one type of chemical.
- All fresh food should be washed well in cold running water. Peel as needed. Soap should not be used in washing fresh food. Fresh produce is very porous, and the detergent will be absorbed, and the effects of detergent on the body are untested.
- People should be aware of food regulations, and they should complain to their local authorities if stores are selling foods with expired dates, etc.

In this modern age, it would be nice to feel the food supply is safe, but one need only to follow the news to understand that all people must take responsibility for safeguarding what they eat.

THE CHALLENGE OF AUTOIMMUNE DISEASES

To date, more than 80 autoimmune diseases have been identified. While some of them, such as rheumatoid arthritis, are relatively common, others are quite rare. Autoimmune diseases affect approximately 5 to 8 percent of the U.S. population, primarily women.

The number of autoimmune diseases is increasing, and scientists do not yet know why nor do they know exactly how to fight some of them. Unlocking this mystery could help improve the quality of life for a large number of people throughout the world. Because these are chronic illnesses, finding cures would greatly decrease medical costs for individuals and society.

The immune system is the body's means of protection against microorganisms and other foreign substances. It is the body's mechanism for fighting disease. In autoimmune diseases, what occurs is that the body turns against itself. Sometimes it is organ specific. However, with a disease such as systemic lupus erythematosus (SLE), any or multiple organs may be under attack when a flare-up occurs.

Over a lifetime, the immune system develops an ability to identify substances and microorganisms that are threatening the body. The vaccination process helps stimulate the body to resist major threats such as measles or diphtheria. The immune system creates antibodies that protect against infectious agents.

With autoimmune diseases, the immune system loses the ability to identify one's own self and may launch an attack against one's own body. The body produces an inappropriate immune response against its own tissues. This causes inflammation and damage.

There is currently no cure for autoimmune disorders, although in rare cases they may disappear on their own. Many people may experience flare-ups and temporary remissions; others have chronic symptoms or a progressive worsening.

Here are just a few of the illnesses categorized as autoimmune diseases by the American Autoimmune Related Diseases Association, Inc:

- Addison's disease (adrenal)
- celiac disease, Crohn's disease, ulcerative colitis (GI tract)
- diabetes mellitus (pancreas islets) type 1
- Hashimoto's thyroiditis, Graves' disease (thyroid)
- Goodpasture's syndrome (lungs, kidneys)
- Guillain-Barré syndrome (nervous system)
- lupus (systemic lupus erythematosus) (skin, joints, kidneys, heart, brain, red blood cells, other)

- multiple sclerosis (scientists debate whether this is an autoimmune disease)
- rheumatoid arthritis (RA) and juvenile RA (joints; less commonly lung, skin)
- scleroderma (skin, intestine, less commonly lung)
- Sjögren's syndrome (salivary glands, tear glands, joints)
- systemic autoimmune diseases localized autoimmune diseases
- Wegener's granulomatosis (blood vessels, sinuses, lungs, kidneys)

Autoimmune disorders are diagnosed, evaluated, and monitored through a combination of autoantibody blood tests, blood tests to measure inflammation and organ function, clinical presentation, and through nonlaboratory examinations such as X-rays.

Treatment of autoimmune disorders is tailored to the individual and may change over time. The goal is to relieve symptoms, minimize organ and tissue damage, and preserve organ function. Research into the causes of autoimmune diseases is focusing on the molecular and cellular levels, and scientists are examining the role inflammation plays in many of the symptoms of the multitude of autoimmune diseases. This should help with creating targeted treatments. The majority of immunologists think that understanding genetic links within the illnesses will eventually permit genetically tailored medicines that should help control—or one day cure—these various puzzling diseases.

CONCLUSION

It seems so simple—eat right and exercise. No matter how far advanced medical science becomes, the bottom line is that the better care an individual takes, the better the chance of a long healthy life. In the meantime, scientists are working hard to lessen the discomfort and duration of colds and flu, and over time they hope to unravel the mysteries of autoimmune diseases, an increasing and devastating issue for modern society.

8

A Medical Visit of Tomorrow

One undeniable change in the current medical system is that an increasing amount of responsibility is being placed on the patient. If a person is diagnosed with a somewhat complex illness, then the primary doctor may refer them to several different specialists, each of whom will provide an opinion. In the long run, the patient is expected to make the choice as to the best option to follow. This is ironic since medical science is becoming more and more complex, yet the ultimate choices are left to relatively uninformed individuals.

Another recent trend in medicine segues with a development in society. As health care costs escalate, people are accessing more and more information through the Internet. Why not visit the doctor virtually? This chapter will note that this trend is indeed happening and will explain what people might encounter. While the emphasis of this volume has been to look forward and consider the future, there are a few areas that have not been mentioned. This chapter will introduce some of the advances in medicine that have not yet received the attention they deserve.

A VIRTUAL DOCTOR VISIT

More and more doctors are using e-mail to respond to patients. If a question is simple, an e-mail response may be sufficient. However, if your child is sick or your mother has just had a stroke, e-mail would definitely be useless.

Virtual doctor visits are just beginning to occur within certain medical practices, but it is almost certainly a trend that will grow. The process begins with the patient visiting the doctor's Web site. The patient would view menu options for services, such as requesting a refill or asking for a referral or checking to see if test results are in. For slightly more complex medical problems, the patient would pick a main symptom, such as back pain, and the system would ask a series of questions.

Doctors appreciate that Web visits are fully documented. This can be helpful if a patient complains about results, who has not complied with the medical advice that was given. Online visits are perfect for simple medical issues, but serious symptoms would call for an in-person visit.

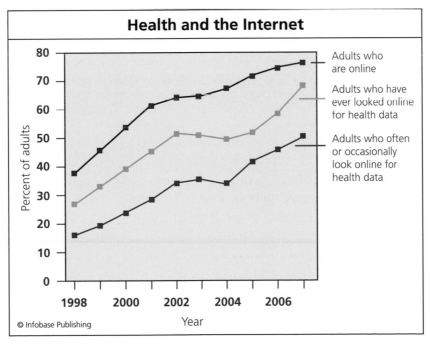

Today, an increasing number of people check out health symptoms on the Internet. This can provide them with more knowledge, but also leads to misdiagnoses. This can result in stress and fear.

CYBERCHONDRIA—FEARS GROW

One problem with Internet access for patients is the fear factor. In 2000, someone actually came up with the term *cyberchondria* to address those who used the Internet and jumped to dire conclusions about their illnesses—the woman who went online to check out her headache may convince herself she has a brain tumor. This has been significant enough that in 2008 Microsoft released a study about health-related Web searches.

Microsoft's rationale for their study was to develop a way to help search engines be more helpful personal advisers. The researcher in charge of the project noted that the syndrome is much like what medical students experience (second-year syndrome or medical schoolitis.) The person experiencing a headache should not be led to the most frightening possibilities first. As students begin to read more deeply about specific illnesses, it is very common to sense similar symptoms. For the Internet to be a helpful tool, Web designers need to find ways to lead people to the less alarming symptoms first.

THE PATIENT, THE EXPERT

Medicine has been forever changed by the ability of people to explore their own medical options and unite with others who have experienced similar issues or illnesses. Called citizen scientists, many are parents of very sick children who feel desperate. They join together to form special interest groups for lobbying, share advice, comment on the most helpful doctors, or exchange news alerts on the disease or medicine. Whether the issue is a genetic abnormality or a child with a cleft palate, people are finding ways to do their own research and reach their own conclusions.

It was actually a citizen activist campaign that brought about the removal of thimerosal (the mercury preservative) from vaccines. While there is a lot of bogus information on the Internet, similar to the patent medicine salesmen of a century ago, there is also a spirit of community.

When it comes to medical problems, there is no replacement for professional guidance. Today, major studies and evidence-based medicine provide doctors with additional information in the race for cures. In addition, a well-informed citizenry strengthens the medical profession's ability to help. The patient who arrives with a detailed list of symptoms and asks good questions and then follows the advice given (whether filling a prescription or starting to exercise) will fare better than those people who are not paying attention. Some patients, on learning that their particular disease or condition needs further study, are pitching in to raise money for research, and this provides scientists with the ability to learn more.

ADVANCES IN MEDICINE

Earlier chapters in this book have discussed progress in the world of medicine—from the increasing number of successful organ transplants to nanomedicine. The following are other areas where progress is growing.

■ Edible vaccines. Vaccines or other medicines may be added to foods. Vaccines have miraculously ended the death toll of six devastating illnesses: diphtheria, pertussis (whooping cough), polio, measles, tetanus, and tuberculosis (which formerly killed about 2 million people each year). A worldwide effort to vaccinate as many children as possible has reached as much as 80 percent of the world's young population. Scientists have puzzled over ways to reach more people and have felt that if a food plant could be genetically altered to contain a vaccine children could get the vaccine without the difficulty of sending and storing and injecting vaccines that usually require refrigeration. Some research has also shown that certain food vaccines might suppress autoimmunity. Autoimmune diseases (discussed in chapter 7) currently have no cure.

Scientist in a lab *(USDA)*

- Programmable pills. In early 2009, a prototype of an intelligent pill was being tested. The idea behind it was that the pill could be programmed to travel to a specific part of the body where the medication it carried would be deposited. For example, aspirin is taken orally and is not site specific. That is why the person who takes too much aspirin may have problems with their stomach. This pill would provide a successful journey so that the medicine could be used where it was needed. This would mean a smaller dose of the drug would be needed with fewer side effects.

- Automated anesthesia. Researchers at McGill University in Montreal have created an automated system to deliver anesthesia and to monitor all of a patient's vital signs, recalibrating the drug concentration based on the readings the automated system is given.

- More advanced prosthetics including mechanical organs. As prosthetic hands are created that can take commands from a person's nervous system and pick up a penny with an acceptable looking artificial hand, researchers are also working to create mechanical hearts, livers, and lungs so that the ceaseless need for transplant organs might be reduced.

■ More advanced surgical techniques. Robotic surgery is changing what surgeons can do, and surgeons are accomplishing amazing things with lasers also. Lasers are used to cut, vaporize, or coagulate tissues with little or no damage to surrounding areas. They are also beneficial in reaching otherwise inaccessible locations with less trauma, bleeding, and scarring. In many cases, surgery with lasers results in reduced postoperative pain, less need for medications, shorter hospitalizations, and quicker returns to routine activities.

There are new discoveries being made every day. Diagnostics, surgery, technology, and pharmaceutical remedies are all still very young fields where progress is constantly being made. There is no telling what will be learned in the next few years.

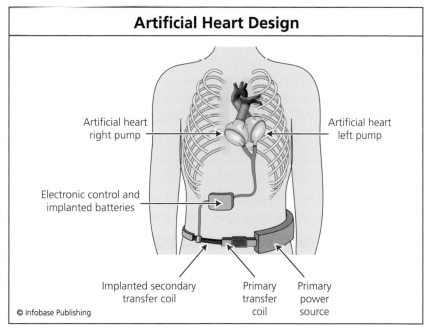

Artificial Heart Design

Artificial heart right pump

Artificial heart left pump

Electronic control and implanted batteries

Implanted secondary transfer coil

Primary transfer coil

Primary power source

© Infobase Publishing

Because human hearts are not always easy to come by, scientists are always looking for ways to create workable, mechanical systems that could be implanted simply and easily.

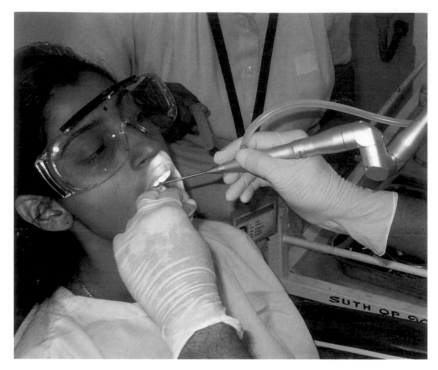

Assistant professor of mechanical engineering Adela Ben-Yakar at the University of Texas at Austin has developed a laser microscalpel that destroys a single cell while leaving nearby cells intact, which could improve the precision of surgeries for cancer, epilepsy, and other diseases. *(Cockrell School of Engineering, University of Texas at Austin)*

NOT ALL ANSWERS ARE MEDICAL

Most of the information in this book has concerned systemic illnesses and their cures. However, some medical conditions affect only vision and can be just as debilitating as a systemic illness. While physicians are working hard to cure diseases ranging from retinal problems to macular degeneration, the true progress that is helping patients is not medical. The advances have come in technology.

People with vision impairment are benefiting from developments that began in the video games industry. All these games have provided medical science with some amazing tools to help people with low vision, and there is no reason that this pace will not continue. Scott Adams, the creator of Dilbert cartoons, has been quoted as saying, "Pretty soon our clothes will be smarter than us."

One day a machine may even be able to run the eye. Under exploration is a retinal chip that might eventually be implanted in the eye to mimic basic photoreceptor cell function. This might help people continue to get around. Two forms of research are underway. One involves placing a retina implant chip over the photoreceptors at the back of the eye (subretinal implant). The other is to put a retinal chip over the ganglion cell layer of the retina (the ganglia are the cells that carry signals from the photoreceptors to the optic nerve). Others are working on a totally artificial retina. These following technological developments can help all who are in any way visually impaired.

- Navigational technologies are getting smaller, cheaper, and more powerful; this will happen exponentially. The prototype wearable computers are now large, expensive, and not powerful enough to fulfill our needs. However, wearable computing or wearable appliances (head mounted, wrist or waist mounted) are becoming less noticeable (blending with natural clothing), cheap enough so that inventors and consumers can buy it from store shelves, and powerful enough so that many needs are addressed, maybe even without a professional to teach the use of it.
- Mobility specialists will be glad when rooms, desk tops, the inside of cars, sidewalks, intersections, hallways, etc., become smart. Wearable computers will then interface, interrelate, and communicate with smart spaces. For example, clothing may one day learn from the intersections where you are and know what to do to help.
- Wearable computers will connect us to other people via telephone, pager, or e-mail access; people will only be out of contact if they really want to be.

- One day, people may buy vests or belts that would vibrate to indicate clear path and be available for the deaf and blind user.
- Devices may exist that could have face recognition and analytical software for sensory enhancement capabilities like digital magnification of video scenes or audio amplification, night vision, or ultraviolet and/or infrared vision.

The fact that many of these devices are being made for the general population is wonderful because this will lead to a faster drop in prices.

CONCLUSION

The 21st century is barely underway, but it is very clear that amazing progress awaits society. From scientific advances that contribute to medical progress to technological gains that are implemented to improve the lifestyles of the disabled or impaired, humankind is poised to make great gains.

However, environmental pollution—including medical waste and the concern over nanotechnology waste—are very real concerns, and science, governments, and individuals need to calculate how to manage these growing problems.

But amazingly the real secret to good health lies not in science or technology. It resides with the individual, who needs to take responsibility for exercising and making healthy choices at mealtimes.

1940s	The entity that was to become the CDC was established as a center for studying epidemiology.
1950s	Dancing cats in Japan led to identification of mercury's potential to poison.
1960s	Stem cells first identified.
1964	Release of first surgeon general's report on smoking and health.
1965	Creation of the Medicare and Medicaid programs, making health care available to the elderly and some of the poor.
1966	International smallpox eradication program was established, led by the U.S. Public Health Service; smallpox was eradicated worldwide by 1977.
1970	Provision for the U.S. Occupational Safety and Health Administration to monitor job safety and job-related illnesses.
1978	Birth of first test tube baby.
1981	AIDS was identified; in 1984, HIV was identified as the virus that caused it.
1984	National Organ Transplantation Act signed into law.
1993	The vaccines for children program is established, providing free immunizations to all children of low-income families.
1997	Creation of the State Children's Health Insurance Program.
	Dolly the sheep was cloned.
2002	The Office of Public Health Emergency Preparedness is created.

2003   Medicare Prescription Drug Improvement and
                           Modernization Act of 2003 is enacted, adding
                           a prescription drug benefit.

                           Carlo Urbani of Médecins Sans Frontières
                           alerted the World Health Organization to the
                           threat of the SARS virus, triggering the most
                           effective response to an epidemic in history,
                           but Urbani dies from the disease in less than a
                           month (March 29, 2003).

2005–06   The world was on heightened alert over avian
                           flu. The first cases of avian flu were identified
                           as early as 1997, and while there is still con-
                           cern that avian flu may become a pandemic,
                           fear was particularly strong in 2005–06.

2007   World's first test tube baby gives birth to natu-
                           rally conceived baby on December 20, 2006.

2009   Barack Obama assumes the presidency, having
                           made promises to loosen the restrictions on
                           stem cell research and to reform health care.

ampule a hermetically sealed bulbous glass vessel that is used to hold a solution for hypodermic injection

anencephalic a congenital absence of all or a major part of the brain

antibiotic a substance produced by or semisynthetic substance derived from a microorganism and able to dilute solution to inhibit or kill another microorganism

autoimmune of, relating to, or caused by the autoantibodies or T cells that attack molecules, cells, or tissues of the organism producing them

beneficence the quality or state of doing a good thing

blastocyst the modified blastula of a placental mammal having an outer layer composed of the trophoblast

chromosome any of the rod-shaped or threadlike DNA-containing structures of cellular organisms that are located in the nucleus of eukaryotes, are usually ring-shaped in prokaryotes (as bacteria), and contain all or most of the genes of the organism; also: the genetic material of a virus

clone a. an aggregate of genetically identical cells or organisms asexually produced by a single progenitor cell or organism; b. an individual grown from a single somatic cell or cell nucleus and genetically identical to it

cortisone a glucocorticoid $C_{21}H_{28}O_5$ of the adrenal cortex used in synthetic form especially as an antiinflammatory agent

culture the act or process of cultivating living material (as bacteria or viruses) in prepared nutrient media; also the product of such cultivation

cytotoxic of or relating to a cytotoxin, toxic to cells

embryo a. vertebrate at any stage of development prior to birth or hatching; b. an animal in the early stages of growth and differentiation that is characterized by cleavage, the laying down of fundamental tissues, and the formation of primitive organs and organ systems; especially the developing human individual from the time of implantation to the end of the eighth week after conception

epidemic   affecting or tending to affect a disproportionately large number of individuals within a population, community, or region at the same time

epidemiology   a branch of medical science that deals with the incidence, distribution, and control of disease in a population

eugenics   a science that deals with the attempted improvement of hereditary qualities of a race or breed, as by control of human mating

genetics   relating to or determined by the origin, development, or causal antecedents of something

graft   to implant (living tissue) surgically

hemorrhage   a copious discharge of blood from the blood vessels

immunity   a quality or state of being immune; especially a condition of being able to resist a particular disease through preventing development of a pathogenic microorganism or by counteracting the effects of its products

immunology   a science that deals with the immune system and the cell-mediated and humoral aspects of immunity and immune responses

immunosuppressive   suppression (as by drugs) of natural immune responses

in utero   literally in the uterus; before birth

in vitro fertilization (IVR)   fertilization of an egg in a laboratory dish or test tube; specifically fertilization by mixing sperm with eggs surgically removed from an ovary followed by uterine implantation of one or more of the resulting fertilized eggs

maleficence   a harmful or evil act

mutation   a significant and basic alteration, change

palliative   serving to soothe or relieve

pandemic   occurring over a wide geographic area and affecting an exceptionally high proportion of the population

placebo   a usually pharmacologically inert preparation prescribed more for the mental relief of the patient than for its actual effect on a disorder

pluripotent   capable of affecting more than one organ or tissue

psychotropic   acting on the mind; usually referring to a class of drugs that affect the central nervous system

radon   a heavy radioactive gaseous element formed by the decay of radium

scourge   an entity that inflicts pain or punishment

totipotent   having the potential for developing in various specialized ways in response to external or internal stimuli

vaccine   a preparation of killed microorganisms, living attenuated organisms, or living fully virulent organisms that is administered to produce or artificially increase immunity to a particular disease

watershed   a region or area bounded peripherally by a divide and draining ultimately to a particular watercourse or body of water

FURTHER RESOURCES

ABOUT SCIENCE AND HISTORY

Diamond, Jared. *Guns, Germs, and Steel: The Fates of Human Societies.* New York: W. W. Norton, 1999. Diamond places the development of human society in context, which is vital to understanding the development of medicine.

Hazen, Robert M., and James Trefil. *Science Matters: Achieving Scientific Literacy.* New York: Doubleday, 1991. A clear and readable overview of scientific principles and how they apply in today's world, including the world of medicine.

Internet History of Science Sourcebook. Available online. URL: http://www.fordham.edu/halsall/science/sciencesbook.html. Accessed July 9, 2008. A rich resource of links related to every era of science history, broken down by disciplines, and exploring philosophical and ethical issues relevant to science and science history.

Lindberg, David C. *The Beginnings of Western Science,* 2nd ed. Chicago: University of Chicago Press, 2007. A helpful explanation of the beginning of science and scientific thought. Though the emphasis is on science in general, there is a chapter on Greek and Roman medicine as well as medicine in medieval times.

Roberts, J. M. *A Short History of the World.* Oxford: Oxford University Press, 1993. This helps place medical developments in context with world events.

Silver, Brian L. *The Ascent of Science.* New York: Oxford University Press, 1998. A sweeping overview of the history of science from the Renaissance to the present.

Spangenburg, Ray, and Diane Kit Moser. *Science Frontiers: 1946 to the Present,* rev. ed. New York: Facts On File, 2004. A highly readable book with key chapters on some of the most significant developments in medicine.

ABOUT THE HISTORY OF MEDICINE

Ackerknecht, Erwin H., M.D. *A Short History of Medicine,* rev. ed. Baltimore, Md.: Johns Hopkins University, 1968. While there have been many new discoveries since Ackerknecht last updated this book, his contributions are still important as they help the modern researcher better understand when certain discoveries were made and how viewpoints have changed over time.

American Medical Association. Available online. URL: http://www. ama-assn.org/. Accessed March 16, 2009. The AMA offers a great deal of interesting information, from recommendations on reforming the health care insurance system to providing ideas for healthier lifestyles.

CDC National Center for Chronic Disease Prevention and Health Promotion. Available online. URL: http://www.cdc.gov/excite/. Accessed March 2, 2009. This is a government Web site dedicated to the teaching of public health and epidemiology.

Centers for Disease Control and Prevention. Available online. URL: http://www.cdc.gov/. Accessed March 1, 2009. The CDC has assembled an amazing Web site featuring basic health topics ranging from on-the-job safety to environmental health. There are also complete alphabetical listings for every illness and health topic imaginable: from a complete description of German measles (rubella) to information on genetic counseling. Because it is government-sponsored, it is both reliable and up-to-date.

Clendening, Logan, ed. *Source Book of Medical History.* New York: Dover Publications, 1942. Clendening has collected excerpts from medical writings from as early as the time of the Egyptian papyri, making this a very valuable reference work.

Dary, David. *Frontier Medicine: From the Atlantic to the Pacific 1492– 1941.* New York: Knopf, 2008. This is a new book that has been very well reviewed. Dary outlines the medical practices in the United States from 1492 forward.

Davies, Gill, ed. *Timetables of Medicine.* New York: Black Dog & Leventhal, 2000. An easy-to-assess chart/time line of medicine with overviews of each period and sidebars on key people and developments in medicine.

Dittrick Medical History Center at Case Western Reserve. Available online. URL: http://www.cwru.edu/artsci/dittrick/site2/. This site provides helpful links to medical museum Web sites. Accessed March 8, 2009.

Duffin, Jacalyn. *History of Medicine.* Toronto: University of Toronto Press, 1999. Though the book is written by only one author, each chapter focuses on the history of a single aspect of medicine, such as surgery or pharmacology. It is a helpful reference book.

Gawande, Atul. *Better: A Surgeon's Notes on Performance.* New York: Henry Holt & Company, 2007. Gawande provides an up-to-date account of the challenges of modern-day medicine, particularly matters concerning surgery.

Groopman, Jerome, M.D. *How Doctors Think.* Boston: Houghton Mifflin, 2007. Dr. Groopman explains how doctors are trained for decision-making and how they go about making an assessment of an illness; most of the time they reach the correct conclusion, but this book helps readers understand what happens when they fail.

Kennedy, Michael T., M.D., FACS. *A Brief History of Disease, Science, and Medicine.* Mission Viejo, Calif.: Asklepiad Press, 2004. Michael Kennedy was a vascular surgeon and now teaches first- and second-year medical students an introduction to clinical medicine at the University of Southern California. The book started as a series of his lectures, but he has woven the material together to offer a cohesive overview of medicine.

Loudon, Irvine, ed. *Western Medicine: An Illustrated History.* Oxford: Oxford University Press, 1997. A variety of experts contribute chapters to this book that covers medicine from Hippocrates through the 20th century.

Magner, Lois N. *A History of Medicine.* Boca Raton, Fla.: Taylor & Francis Group, 2005. An excellent overview of the world of medicine from paleopathology to microbiology.

Medical Discoveries. Available online. URL: http://www.discoveriesinmedicine.com/. Accessed February 26, 2009. This Web site provides an alphabetical resource with biographies and other information about important medical milestones.

National Human Genome Research Institute. Available online. URL: http://www.genome.gov/. Accessed February 1, 2009. This is a National Institutes of Health government-sponsored site to provide the public with information about all aspects of research concerning the human genome.

Porter, Roy, ed. *The Cambridge Illustrated History of Medicine.* Cambridge, Mass.: Cambridge University Press, 2001. In essays written by experts in the field, this illustrated history traces the evolution of medicine from the contributions made by early Greek physicians through the Renaissance, scientific revolution, and 19th and 20th centuries up to current advances. Sidebars cover parallel social or political events and certain diseases.

———. *The Greatest Benefit to Mankind: A Medical History of Humanity.* New York: W. W. Norton, 1997. Over his lifetime, Porter wrote a great amount about the history of medicine, and this book is a valuable and readable detailed description of the history of medicine.

Rosen, George. *A History of Public Health, Expanded Edition.* Baltimore: Johns Hopkins University Press, 1993. While serious public health programs did not get underway until the 19th century, Rosen begins with some of the successes and failures of much earlier times.

Simmons, John Galbraith. *Doctors & Discoveries.* Boston: Houghton Mifflin Company, 2002. This book focuses on the personalities behind the discoveries and adds a human dimension to the history of medicine.

Spangenburg, Ray, and Diane Kit Moser. *Disease Fighters Since 1950.* New York: Facts On File, 1996. A highly readable book with key chapters on the discovery of AIDS as well as the beginning of the movement toward improving world health.

Starr, Paul. *The Social Transformation of American Medicine.* New York: Perseus, 1982. The book puts in perspective the changes in the American medical system and how they came about.

United States National Library, National Institutes of Health. Available online. URL: http://www.nlm.nih.gov/hmd/. Accessed February 23, 2009. A reliable resource for online information pertaining to the history of medicine.

OTHER RESOURCES

Collins, Gail. *America's Women: 400 Years of Dolls, Drudges, Helpmates, and Heroines.* New York: William Morrow, 2003. Collins' book contains some very interesting stories about women and their roles in health care during the early days of America.

Drexler, Eric. *Engines of Creation: The Coming Era of Nanotechnology.* New York: Anchor Books, 1987. This is a dated book on nanotechnology but Drexler is considered the "father of nanotechnology," so it is helpful to anyone trying to understand the underpinnings of the field.

National Nanotechnology Initiative. Available online. URL: http://www.nano.gov/. Accessed March 3, 2009. Another government-sponsored Web site, this one provides current information on developments in nanotechnology.

INDEX

Note: Page numbers in *italic* refer to illustrations; *m* indicates a map; *t* indicates a table.

A

acquired immunodeficiency syndrome (AIDS). *See* HIV/AIDS
Adams, Scott 141
Addressing Obesity through Commercial Health Plans 120–121
adult (somatic) stem cells 72, 79
African Americans, syphilis study 96–98
aging 11, *123m*
agriculture 39–41
Agriculture, Department of 125–126
AIDS. *See* HIV/AIDS
amalgams 27, 29
American Academy of Pediatrics 93
American Autoimmune Related Diseases Association 132–133
American Physical Society 83
American Society for Microbiology 38
anencephaly 92–93
anesthesia, automated 138
animal-human diseases
    climate change and 65–67
    conservation medicine 61–64
    HIV/AIDS 8, 50–51
    human behavior and 47–48
    influenza. *See* influenza
    SARS *48*, 48–50, *49m*
antibodies 125, 132
antiviral medications (Tamiflu, Relenza) 58, 61

artificial hearts *139*
asbestos 30–31, 33–36, *34, 35*
asbestosis 33–34
aspirin 95–96
Associated Press (AP) 44
asthma 38, 40–41
atoms 85
autism 29
autoimmune diseases 131–133
automated anesthesia 138
automated external defibrillators (AEDs) 10
avian flu (H5N1) *51*, 51–57, *52, 54m, 56*, 62, 63

B

babesia 65
Baby K 92–93
bats 49
Becker, Andy 71
Ben-Yakar, Adela *140*
Billingham, Rupert 13
Binnig, Gerd 83–84
bioengineered foods. *See* genetically altered foods
bird flu. *See* avian flu (H5N1)
blastocysts. *See* embryos (blastocysts)
Blue Cross and Blue Shield 105
blue mass pills 28
bone breaks 87
bovine tuberculosis 67
Brigham and Women's Hospital 14–15, 16
Brown, Louise 18, 19–20
Burnham Institute for Medical Research 123
Bush, George W. 30, 73
Buxtun, Peter 98